The Research of Sustainability Assessment for Residential District in China

中国大型城镇住区发展的可持续性评估工具研究

赵博阳 张小弸 著

天津大学出版社
TIANJIN UNIVERSITY PRESS

图书在版编目（CIP）数据

中国大型城镇住区发展的可持续性评估工具研究 / 赵博阳, 张小珊著. -- 天津 : 天津大学出版社, 2024. 10. -- ISBN 978-7-5618-7758-6

Ⅰ. TU984.12

中国国家版本馆 CIP 数据核字第 20248KW403 号

中国大型城镇住区发展的可持续性评估工具研究 | ZHONGGUO DAXING CHENGZHEN ZHUQU FAZHAN DE KECHIXUXING PINGGU GONGJU YANJIU

出版发行	天津大学出版社
地　　址	天津市卫津路 92 号天津大学内（邮编：300072）
电　　话	发行部：022-27403647
网　　址	publish.tju.edu.cn
印　　刷	北京虎彩文化传播有限公司
经　　销	全国各地新华书店
开　　本	787 mm × 1 092 mm　1/16
印　　张	8.5
字　　数	235 千
版　　次	2024 年 10 月第 1 版
印　　次	2024 年 10 月第 1 次
定　　价	45.00 元

凡购本书，如有缺页、倒页、脱页等质量问题，烦请与我社发行部门联系调换
版权所有　　侵权必究

中国大型城镇住区发展的可持续性评估工具研究

The Research of Sustainability Assessment for Residential District in China

Abstract

This research attempts to make a better understanding of the concept of sustainable urban residential district assessment and its role in sustainable development in China. Moreover, the research tries to establish a framework of feasible and effective sustainable residential district assessment items for China. And then, based on the framework of this research, a Sustainable Residential District (SRD) Rating System (a computer program) to measure the performance on the sustainability issues will be developed, and thus to promote the sustainable development of the urbanization of China.

Over the past decades, the importance of sustainable development has been recognized by the whole world. Meanwhile, in the building industry many green building assessment tools such as BREEAM, LEED, CASBEE and others were established in many developed countries. Now, these assessment tools have developed various versions for different building styles and scales and the assessment tools for residential area development are the latest versions of BREEAM, LEED and CASBEE.

On the other hand, with the fast urbanization, a lot of big scale residential developments such as new town projects emerge in China in recent years, which creates an urgent demand for sustainability assessment tools on the residential area scale.

This research firstly executed a systemic analysis on the Chinese urban planning system and residential development process and an integrated study on the state of sustainable development and green building in China. The research identified the necessity of sustainability assessment on the area scale and discussed the possibility and potential benefits of the residential district as the assessment scale.

摘 要

本研究致力于深化对可持续城市居住区评估概念的理解,并探讨其在中国可持续发展战略中的关键作用。本研究尝试构建一个具有可操作性和实效性的可持续居住区评估框架。在这一框架基础上,研究开发用于量化评估居住区在可持续发展方面表现的可持续居住区(SRD)发展评级系统工具,希望以此来推动中国城市化的绿色进程。

近几十年来,可持续发展的重要性已在全球范围内得到广泛认同。在建筑行业中,许多发达国家已经建立了诸如 BREEAM、LEED 和 CASBEE 等绿色建筑评估工具。这些评估工具已开发出针对不同建筑类型的多样化版本,其中,针对住区的评估工具是这些系统近年来推出的较新版本。与此同时,随着中国城市化进程的快速推进,近年来涌现出大量大型城镇住宅开发项目,这凸显出对住区可持续性评估工具的迫切需求。

在这一背景下,本书首先系统分析了中国城市规划体系以及住宅发展历程,并综合研究了中国的可持续发展现状和绿色建筑实践情况。通过这一分析,明确了在城镇片区尺度上进行可持续性评估的必要性,并探讨了以大型住区作为评估尺度的可行性和潜在效益。我们对现有的 3 种居住区尺度评估工具——LEED for Neighborhood、BREEAM Communities 和 CASBEE for Urban Development 进行了深入比较和分析。在充分考虑我国本土实际的基础上,我们构建了一个初步的评估项目框架,并通过问卷等方式进一步优化并确定了评估框架,同时借助层次分析法为最终的评价项目建立了较为科学的权重体系。

Then the research made a comparison and deep analysis of the three existing assessment tools for residential area scale, i.e., LEED for Neighborhood Development, BREEAM Communities and CASBEE for Urban Development, and built an initial framework of assessment items respecting the local context of China. And then the initial framework was improved and determined through questionnaires. With the help of the Analytic Hierarchy Process (AHP) methodology, the weighting system of the final assessment items was established.

The outcome of this research established the suggested Sustainable Residential District (SRD) Rating System (a computer program). With this tool, four residential district developments in China were assessed and the results were used to test and prove its practicability and usefulness. The suggested assessment tool can assist the participants of residential district development in China to make a general and clearer view of the residential district projects on the sustainability issues and further to improve their sustainability performances.

本研究的重要成果之一是建立了可持续居住区（SRD）发展评级系统这一评估工具。为了验证该工具的实用性和有效性，我们选择了中国的4个大型城镇住区进行评估测试。结果表明，该评估工具能够帮助中国住宅区开发的各方参与者更全面、更清晰地了解住宅区项目的可持续性状况，从而有针对性地提升其可持续性表现。

Table of Contents

Chapter One: Introduction ... 1
  1.1 Research context ... 3
  1.2 Research purpose and objectives ... 6
  1.3 Research structure ... 7

Chapter Two: Literature Review ... 11
  2.1 Sustainable development and sustainability ... 13
  2.2 The research of green building and assessment ... 15
    2.2.1 The concept of green building ... 15
    2.2.2 Green building assessment tools and rating systems ... 16
  2.3 Representatives of international green building assessment systems ... 18
    2.3.1 Great Britain: BREEAM (Building Research Establishment's Environmental Assessment Method) ... 18
    2.3.2 United States of America: LEED (Leadership in Energy and Environmental Design) ... 20
    2.3.3 Japan: CASBEE (Comprehensive Assessment System for Building Environmental Efficiency) ... 23
    2.3.4 Italy: ITACA Protocol ... 25
  2.4 Research of sustainable residential area and assessment ... 26
    2.4.1 Concept of sustainable residential area ... 26
    2.4.2 Attempts of sustainable residential area practice ... 27

目录

第1章 绪论 1
  1.1 研究背景 3
  1.2 研究目的与目标 6
  1.3 研究框架 7

第2章 文献综述 11
  2.1 可持续发展与可持续性 13
  2.2 绿色建筑及其评价研究 15
    2.2.1 绿色建筑的概念 15
    2.2.2 绿色建筑评估工具和评级系统 16
  2.3 国际绿色建筑评估系统的代表 18
    2.3.1 英国：英国建筑研究院环境评估方法（BREEAM） 18
    2.3.2 美国：美国能源与环境设计先导评价标准（LEED） 20
    2.3.3 日本：日本建筑物综合环境性能评价体系（CASBEE） 23
    2.3.4 意大利：意大利环境与采购活动的创新性与透明度协议（ITACA Protocol） 25
  2.4 可持续居住区研究 26
    2.4.1 可持续居住区的概念 26
    2.4.2 可持续社区的设计实践 27

2.5 Representatives of existing sustainable residential area assessment tools 32
    2.5.1 From green building assessment to residential area assessment 32
    2.5.2 BREEAM Communities 32
    2.5.3 LEED for Neighborhood Development 34
    2.5.4 CASBEE for Urban Development 36

Chapter Three: Assessment Scale and Platform Study of Sustainable Residential District Assessment in China 39
  3.1 Sustainable development and green building in China 40
    3.1.1 Status quo of sustainable development and green building in China 40
    3.1.2 Status quo of green building assessment in China 43
  3.2 Research of the assessment scale and platform in China 46
    3.2.1 Introduction of Chinese urban planning system and residential development process 46
    3.2.2 Study of the suitable scale and platform of sustainable residential district assessment 49

Chapter Four: Establishment Process of Sustainable Residential District Assessment Tool 51
  4.1 Introduction of research design 53
  4.2 Preparation of the initial framework of assessment items and questionnaire 56
    4.2.1 Introduction 56
    4.2.2 The assessment items comparison and analysis of existing sustainability assessment tools for area scale in developed countries 58
    4.2.3 The initial framework of assessment items 73
  4.3 Establishment of the final framework of assessment items 77
    4.3.1 Introduction 77
    4.3.2 Sampling procedure and Questionnaire 1 & 2 79
    4.3.3 Result analysis of Questionnaire 1 & 2 85
    4.3.4 Final framework of assessment items 87
  4.4 Weighting coefficient system set-up 90
    4.4.1 Introduction 90
    4.4.2 The methodology of Analytic Hierarchy Process (AHP) 90

2.5 现有可持续居住区评估工具的代表　　32
　　　　2.5.1 从绿色建筑评估到住区评估　　32
　　　　2.5.2 BREEAM 社区　　32
　　　　2.5.3 LEED 社区发展　　34
　　　　2.5.4 CASBEE 城市发展　　36

第3章　我国大型城镇住区可持续性评估的尺度与平台研究　　39
　　3.1 我国的可持续发展与绿色建筑　　40
　　　　3.1.1 我国可持续发展与绿色建筑现状　　40
　　　　3.1.2 我国绿色建筑评估现状　　43
　　3.2 我国的评估范围与评估平台研究　　46
　　　　3.2.1 我国城市规划体系和住宅开发过程介绍　　46
　　　　3.2.2 可持续居住区评价的适宜规模与平台研究　　49

第4章　我国大型城镇住区可持续性评估工具的建立过程　　51
　　4.1 研究设计简介　　53
　　4.2 编制评估指标和调查问卷的初步框架　　56
　　　　4.2.1 初步框架说明　　56
　　　　4.2.2 发达国家现有住区规模可持续性评估工具的评估指标比较分析　　58
　　　　4.2.3 建立评估指标的初步框架　　73
　　4.3 建立评估指标的最终框架　　77
　　　　4.3.1 调查问卷设计说明　　77
　　　　4.3.2 抽样程序和调查问卷1与调查问卷2　　79
　　　　4.3.3 问卷1与问卷2的结果分析　　85
　　　　4.3.4 建立评估指标的最终框架　　87
　　4.4 设置权重系统　　90
　　　　4.4.1 设置权重方法简介　　90
　　　　4.4.2 层次分析法（AHP）　　90

  4.4.3 Developing the hierarchic structure of weighting coefficient system and Questionnaire 3   93

  4.4.4 Result and analysis   94

  4.4.5 Comparison among the categories' weighting coefficient of BREEAM Communities, LEED for Neighborhood Development, CASBEE for Urban Development and SRD rating system   94

 4.5 Sustainable Residential District (SRD) Rating System for China   96

  4.5.1 Introduction   96

  4.5.2 The excel spread sheets of Sustainable Residential District (SRD) Assessment Rating System for China   98

  4.5.3 The technical handbook of Sustainable Residential District (SRD) Assessment Rating System for China   98

Chapter Five: Tests of the Sustainable Residential District (SRD) Rating System for China   103

 5.1 Introduction   105

 5.2 Test 1: The project of Huaming Town   106

 5.3 Test 2: The project of Balitai Town   108

 5.4 Test 3: The project of Shuangkou Town   108

 5.5 Test 4: The project of Gaojiazhuang Town   109

 5.6 Comparison and analysis of the assessment results of the four test cases   109

Chapter Six: Conclusion and Future Direction of the Research   113

 6.1 Introduction   115

 6.2 Research Conclusion   115

 6.3 Research Contribution   117

 6.4 Limitations of the research   118

 6.5 Future direction of the research   118

Bibliography   121

  4.4.3 开发加权系数体系的层次结构和问卷 3 93
  4.4.4 结果与分析 94
  4.4.5 与 BREEAM 社区、邻里发展 LEED、城市发展 CASBEE 和 SRD 评级体系类别权重系数的比较 94
 4.5 中国可持续居住区（SRD）评估系统 96
  4.5.1 评估系统简介 96
  4.5.2 建立我国可持续居住区（SRD）评估体系 98
  4.5.3 编制我国可持续居住区评估系统技术手册 98

第 5 章 我国大型城镇住区可持续性评估工具的应用检验 103
 5.1 本章简介 105
 5.2 案例 1：华明镇项目 106
 5.3 案例 2：八里台镇项目 108
 5.4 案例 3：双口镇项目 108
 5.5 案例 4：高家庄镇项目 109
 5.6 4 个测试案例评估结果的比较与分析 109

第 6 章 研究的结论和未来的研究方向 113
 6.1 本章简介 115
 6.2 研究总结 115
 6.3 主要价值与贡献 117
 6.4 研究的局限性 118
 6.5 未来研究方向 118

参考文献 121

Chapter One: Introduction

第1章 绪论

在全球城市化迅猛发展的背景下，自 20 世纪末开始，我国的城市空间扩张步入了高速发展的轨道。这种城市空间的高速增长不仅有力地推动了经济的发展，同时也引发了一系列社会和环境方面的问题。作为城市中占据主导地位的建筑类型，居住建筑因其巨大的市场需求而大量兴建，推动了我国不同地区大型城镇住区的蓬勃发展。然而，这种快速的发展态势也对住区尺度的可持续评价工具提出了更为迫切的需求，以期在保障经济发展的同时，有效应对社会和环境方面的挑战。

BREEAM Communities、LEED for Neighborhood Development 以及 CASBEE for Urban Development 是目前较为成熟且完善的 3 种社区可持续性评估工具。它们不仅有效地将可持续发展和绿色建筑的基本理念应用于社区尺度的实践中，同时也为本研究提供了宝贵的参考和借鉴。这些评估工具在推动社区可持续发展方面发挥着重要作用，对于未来住区建设具有重要的指导意义。

尽管可持续性评价工具对于我国大规模居住区的发展具有重要意义，然而其重要性却并未得到广泛的认同。我国依然缺乏一个能够有效指导住区发展的通用性评估工具。此外，在住区开发的实践中，大量引入国外的概念、观念、经验和技术，却缺乏本土化的思考、研究和改进，导致这些外来元素与我国的国情产生了显著的冲突。如何将这些元素有效地融入我国的实际环境中，实现其在我国语境下的合作与应用，也成为我国研究者亟待解决的一个重要问题。

因此，针对我国大型城镇住区的发展，迫切需要一个适合的可持续性评价工具来指导实践。本研究致力于构建一个可行且有效的中国大型城镇住区可持续发展评价框架，并在此基础上进一步提出了大型城镇住区可持续性的评价工具，为推动中国城市的可持续发展提供有力支持。

本研究主要通过以下 4 个方面的内容来实现研究目的。

首先，对我国可持续发展与绿色建筑的现状进行了深入调查，全面了解可持续发展与绿色建筑工具开发的背景与现状。通过这一研究为后续的评估工具开发提供坚实的基础。

其次，针对中国可持续居住区评价量表及平台进行了系统研究。进一步探讨并分析我国的土地所有权、规划制度以及房地产开发程序，以期寻找最适合的评估工具规模和平台。这一研究对于确保评估工具的有效性和适用性具有重要意义。

然后，对大规模住宅小区开发可持续性评价工具的研究现状进行了深入了解。在这一过程中，对现在所有的评估工具进行了全面的调查、比较和分析。通过这一步骤，我们期望能够发现现有评估工具的优缺点，为后续的改进和创新提供有力支持。

最后，本研究选取了一系列具有代表性的案例进行测试，以评估其在可持续发展方面的表现。这些案例均具备适当的规模，且近年来在规划、开发和建设方面取得了显著成果，能够代表我国大型城镇住区的可持续发展问题的总体表现水平。通过对测试结果进行比较和分析，我们期望进一步完善评分系统，提高评估工具的准确性和有效性。

本书共分为6章。第1章为绪论部分，详细阐述了研究的背景信息，明确了当前研究现状中所面临的关键问题，进而确定了本研究的目标和研究方向。第2章是文献综述，系统梳理了与"住区评价"相关的全球最新研究成果。该章深入探讨了可持续发展、绿色建筑以及绿色建筑评价等核心概念的起源和发展历程，为后续研究提供了坚实的理论基础。第3章聚焦于我国可持续居住区评价的尺度与平台研究。通过对我国土地所有制、规划制度以及房地产开发程序的深入分析，本章旨在探索适合我国国情的评估工具规模和平台，为确保评估的准确性和有效性提供重要支撑。第4章详细阐述了可持续居住区评价工具的建立过程。本章采用了编制评估指标、抽样和问卷调查、建立权重体系、制定评估表和评级体系等4个步骤的方法论框架，力求构建一个符合中国实际的可持续居住区评级体系。第5章是对中国可持续居住区（SRD）评价体系的实证检验。在这一章中，我们选取了几个具有代表性的案例进行评估，作为对所提出系统的测试。通过对这些评估结果的深入分析，我们期望进一步完善该评价工具，提升其适用性和准确性。第6章总结了本研究的主要发现和结论，归纳了本研究的创新点和意义，指出了研究中存在的局限性，并为未来的进一步研究提供了有益的建议和方向，以期推动该领域的持续发展。

1.1 Research context

Nowadays, 50.5% or 3.5 billion of the people on Earth are living in cities in 2010 and the urban population is still growing, often at the expense of rural areas, making the global population as a whole become more urban and less rural (UN, 2009). Since the World War II, the world's urban population has increased rapidly and in the next few decades a great growth of urban scale will be seen all over the world especially in the developing countries. It can be forecasted that the urban population will grow twice from 2000 to 2030 in Africa and Asia, which means that the accumulated urbanization of these two regions during the last hundreds of years will be doubled in just one single generation (UN, 2009).

Since the Reform and Opening Up which started in the late 1970's, the fast expansion of urban space has become one of the most obvious characters of Chinese urbanization. From 1990 to 2008, the urban construction area in China had expanded from 13,000 km^2 to 36,000 km^2. The elasticity coefficient of urban land (Urban land Growth Rate/Urban Population Growth Rate) increased from 2.13 during 1986−1991 to 2.28, which was much higher than the rational level.

The expansion speed of urban space was prodigious especially in some developed regions. According to the data of MOHURD, in Jiangsu Province, the urban built area increased from 2,119.5 km^2 in 2003 to 2,904.32 km^2 in 2008, increasing almost 37.02% in 5 years; In Zhejiang Province, the urban built area increased from 1,397 km^2 in 2003 to 1,939.09 km^2 in 2008, increasing 38.8% in 5 years; Among these regions, Guangdong Province expanded the fastest, from 2,546.9 km^2 to 4,132.63 km^2, increasing 62.26% in 5 years.

While in the middle regions of China, the expansion of urban space also kept pace with developed regions especially in Hebei, Shandong, Jilin Provinces and Chongqing Municipality City. In Hebei Province, the urban built area was 1,171 km^2 in 2003. While in 2008, the number increased to 1,528.33 km^2, increasing 30.52% in 5 years. In Shandong Province, the number rose from 2,195.4 km^2 in 2003 to 3,261.03 km^2 in 2008. Such a speed was also seen in some undeveloped regions in China. In Yunnan Province, the urban built area increased 51.9%, which means an increase of 213.16 km^2 in 5 years.

From the data above we can see that now China is experiencing a fast urbanization. Urban space is growing at an alarming pace, placing increasing pressure on economic, social and environmental aspects. In fact, the fast urbanization brought similar influence at global scale. Just as Judith Roding, president of the Rockefeller Foundation, said, "With clear and compelling evidence that urban future will be distinct and powerful, we can't go back. We must build new and more resilient physical, economic and social infrastructures for the 21st century, as we came out of the Great Recession."

Therefore, sustainable development was considered as a key to solve the economic, social and environmental problems in the urbanization of 21st century. Sustainable Development is development which meets the needs of the present without compromising the ability of future generations to meet their own needs (WCED, 2003). It underscores the importance of taking a longer-term perspective about the consequences of today's activities, and of global cooperation among countries to reach viable solutions and also ensure that today's development does not exhaust the growth possibilities in the future. It can be used to cover very divergent ideas (Adams,

2001).

In order to guide the building and city by the sustainable strategies and principles, diverse evaluation systems and researches were made in the last 20 years. The evaluation systems firstly appeared in the field of green building. In order to measure how green, or sustainable a building is, a series of green rating systems have been developed, such as Leadership in Energy and Environmental Design (LEED), Building Research Establishment's Environmental Assessment Method (BREEAM), Comprehensive Assessment System for Building Environmental Efficiency (CASBEE), Green Building Tool (GBTool), etc. Then, after the green assessments of individual building were well accepted by the markets, most representative green building assessment systems began to explore a special version for the sustainability assessment of big scale area development. Until now, BREEAM Communities (2009), LEED for Neighborhood Development (2009) and CASBEE for Urban Development (2007) have been the three existing sustainability assessment tools for residential area development. The basic concepts and ideas of sustainable development and green building are well accepted by these systems. These new established tools can measure the sustainability of an area development and with a rating system to determine how sustainable the project assessed is.

Now the environmental assessments are commonly adopted in large-scale development projects. Different from the environment assessments, these sustainable residential area assessment tools have different concepts and roles in the big scale developments. They address the importance of various actions against environmental, economic and social problems and make a comprehensive assessment of the merits and demerits on the environmental, economic and social sustainability of a development.

On the other hand, as a main part of city, the residential area and residential building take a huge percentage of the construction industry. In fact, the residential building construction now is a key element for China to keep the rapid pace of the economic growth. China's rapid economic expansion is driven by the rapid urbanization, by the huge amount of land exchanged and by construction materials and equipment produced for using in buildings. In recent years, there are about 50~100 million m^2 residential building completed in China every year.

Meanwhile, the huge demand and construction of residential building promotes the big scale residential area development in different regions of China. They are often called as New Town or Residential District. In China, the Residential District is defined in the national standard as those residential settlements which are surrounded by arterial roads or natural boundaries and generally have a population of 30,000~50,000 or more. Due to the special land ownership, urban planning system, the development of residential district is one the most important steps of residential development procedure in China, which connects the development of city scale with the construction of building and site (block) scale. However, there has been no sustainability assessment tool for the area scale until now.

1.2 Research purpose and objectives

Now, the significant role of the sustainability assessment tool for big scale residential area development in China has not been widely recognized. Though the importance to conserve environment, to mitigate the environmental deterioration and some other issues of sustainable development have been broadly acknowledged by Chinese governments, estate developer, urban planner and building designer, we still lack a general tool which could guide the urbanization and the residential development in China. In addition, too many foreign concepts, ideas, experiences and techniques were applied in the practices of residential development without local consideration, research and improvement, which produced many conflicts with the context of China. How to make these issues integrated and applied well in the context of China is also a question for the Chinese researchers.

In such condition, China urgently needs to determine a sustainability assessment tool for big scale residential area development. Relevant research work in this field is inadequate. In this context, this research attempts to establish a framework of feasible and effective sustainable residential district assessment items for China. Then, based on the framework this research will develop a suggested Sustainable Residential District (SRD) Rating System (a computer program) to measure the performance on the sustainability issues, and thus to promote the sustainable development in the urbanization of China.

For achieving the main aim of this research, a number of objectives are derived. The main objectives of this research are as follows.

1. The state of Sustainable development and green building in China.

The state of sustainable development and green building in China would be investigated firstly in order to understand the context in which the tool is developed.

2. Assessment scale and platform study of Sustainable Residential District Assessment in China.

Different from those assessment tools for building, the scale and platform are essential for the tool of area development. A further study and analysis of the special ownership of land, planning system and estate development procedure in China would be made to find the best scale and platform for the assessment tool.

3. The state of sustainability assessment tools for big scale residential area development.

This research tries to make a better understanding of the concept and technical details of the sustainability assessment tools for big scale residential area development. All the existing assessment tools should be investigated, compared and analyzed.

4. The performance on the sustainability issues of test cases.

As the test step, several residential district developments will be assessed by the suggested sustainability assessment tool. They all have suitable scales, planned, developed and constructed in recent years, and could represent the general performance level on sustainability issues of residential district developments in China. The results will be compared and analyzed to improve the rating system. In addition, such comparison and analysis would be helpful to understand the state of sustainable residential district development in China and to know how to improve them in the future.

1.3 Research structure

Present research can be divided into six chapters. Fig 1.1 displays the research structure including principal contents of the research. The logic structure and contents in each chapter of the research are briefly introduced as follows.

Chapter One is an introduction for the whole research, which discusses the context and defines the present problems to be solved. It identifies the target, the objectives and the structure of the research.

Chapter Two is literature review which tried to find out the most relevant and up-to-date research in the world. The origin and development history of the "Sustainable Residential Area Assessment", the conception of Sustainable Development, Green Building and Green Building Assessment were studied.

Chapter Three is assessment scale and platform study of Sustainable Residential District Assessment in China. Different from those assessment tools for building, the scale and platform are essential for the success of the assessment tools for area development. This chapter tries to find the best scale and platform for the assessment tool after a further study and analysis of the special ownership of land, planning system and estate development procedure in China.

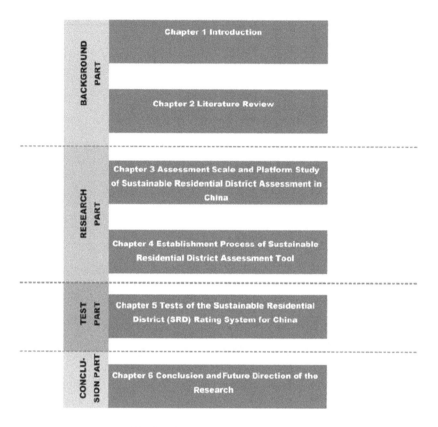

Fig 1.1 Structure of the research

Chapter Four is the establishment process of the Sustainable Residential District Assessment Tool. A four-step method which includes Preparation, Sampling and Questionnaire, Weighting System Set-Up and Assessment Sheets & Rating System Set-Up is adopted by this research to explore the suggested rating system— Sustainable Residential District (SRD) Rating System for China.

Chapter Five is tests of the Sustainable Residential District (SRD) Rating System for China. In this chapter, several residential district projects in China will be assessed as tests for the suggested system. These results will be analyzed to improve the tool.

Chapter Six is conclusion and future direction of the research. This chapter summarizes the main findings and conclusions with the purpose and objects of this research proposed at the beginning. Then, it concludes the innovation and significance of the research, and points out the limitations of the research. It also provides some suggestions and directions for the further research in future.

Chapter Two: Literature Review

第2章 文献综述

"可持续发展"的概念在1987年联合国环境与发展委员会的报告《我们共同的未来》中被首次提出，为人类社会的发展道路提供了新的视角和方向。按照该报告的定义，可持续发展指的是既满足当代人的需要，又不损害后代人满足其需求的发展。换言之，它要求我们在经济、社会和环境3个维度上实现平衡和协调的发展。随着时间的推移，可持续发展理论已经逐渐成为全球大部分国家推动社会进步、经济增长和生态环境保护的重要指导思想。

在建筑领域，绿色建筑的理念于20世纪60年代在国外初见端倪。随着20世纪70年代石油危机的冲击，节能在绿色建筑理念中逐步占据了核心地位，并推动该理念不断深化，最终在20世纪80年代形成了一套相对成熟的建筑节能框架。进入20世纪90年代后，绿色建筑理论进一步得到丰富和完善，逐渐形成了一套涵盖建筑全生命周期的绿色建筑评价及认证体系。其中，由英国建筑研究所开创的BREEAM（英国建筑环境效率评估法）标志着绿色建筑评估体系的诞生，它在全球范围内具有里程碑意义。随后，美国绿色建筑协会、日本建筑物综合环境评价研究委员会等组织，结合各自国家的实际情况，也相继开发了各具特色的绿色建筑评估工具。历经数十年的研究与实践，这些评估工具从最初的定性分析逐步发展为定量评估，其评价指标也从单纯的技术性能拓展到经济、社会和环境等多重维度的综合考量。这些评估体系的广泛应用，对于引导世界各国在建筑实践中融入绿色理念、推动城市的可持续发展起到了举足轻重的作用。

尽管关于可持续居住区的精确定义和概念仍在学术界的探讨之中，但不可忽视的是，全球范围内针对可持续居住区的实验与实践正在如火如荼地进行。世界各地已经涌现出众多杰出的可持续居住区项目，诸如英国的贝丁顿零能源开发社区（BedZED）、德国的弗赖堡沃班社区，以及我国的天津中心生态城等。这些实践案例不仅为城市的可持续发展提供了有力的实证支持，更在很大程度上推动了全球城市可持续建设理念的深化与实践的拓展。

同时，多数具有代表性的绿色建筑评估体系也开始致力于研发专门用于评估住区规模可持续性的工具。当时大型开发项目普遍采用环境评价方法，与环境评估不同，可持续居住区评估在大规模开发中有不同的概念和作用，其更加关注环境、经济和社会问题的重要性，并对环境、经济及社会可持续性的优点和缺点进行了全面评估。目前，BREEAM Communities、LEED for Neighborhood Development以及CASBEE for Urban Development是3种相对完善的社区可持续性评估工具。对这三大评估体系的深入研究，不仅有助于在社区层面上实践可持续发展和绿色建筑的基本理念，同时也为本研究提供了宝贵的参考依据。

2.1 Sustainable development and sustainability

Nowadays, the term of sustainable development can be widely accepted as a means of enhancing quality of life and thus allowing people to live in a much healthier environment and improve social, economic and environmental conditions for present and future generations all over the world. However, the original concept of sustainability can be traced back to the discussion more than 40 years ago. In the 1970s and 1980s, a series of documents and reports on sustainability began to be published in some developed countries. In fact, the movement of sustainable development started then, launched with two documents which have worldwide influences: One is *Limits to Growth* (Meadows, Rangers, et al., 1972) reported by the Club of Rome who presented historical arguments from Thomas Malthus, which were brought into the modern context as "Malthusian disaster" predicted by Malthus in eighteenth century that the fast human population growth will exceed the ecological limits of food productions, which would lead to drastic die-off of human finally. Another document is *Our Common Future* (Brundtland Commission, 1987) reported by the World Commission on Environment and Development (WCED). In this document, sustainable development is firstly defined as "development that meets the needs of the present without compromising the ability of future generations to meet their own needs". At the same time, the document concentrated on the world's economic and environmental problems. It was pointed out that environmental problems were closely tied to inequality of economic and social development. It is indicated that the economic development should not stop while it must change the mode to fit within the limits of nature. The document purposed to find a mode of economic development to protect the environment and ecosystem. It gave a worldwide accepted definition of the term "sustainable development". This combination of sustainability and development tries to reconcile economic growth in the neoclassical tradition with a new concern for environmental protection, recognizing the biophysical "limits to growth" (Meadows et al. 1972) as a constraint to economic development (Adams, 2006).

Since then, sustainable development has gained much more attention from all over the world. In 1992, The United Nations Conference on Environment and Development (UNCED), also known as the Rio Summit which had significant influence on the popularization and development of sustainability, was held in Rio de Janeiro, Brazil. It was emphasized that "sustainable development" is a solution to environmental and social problems with effective measures to prevent environmental degradation and economic and social progress depends critically on the preservation of the natural resource. The Rio Summit is considered as "an important event for the internationalization or globalization of science" (Hibbard, Crutzen, Lambin, et al., 2007), directly or indirectly responsible for subsequent conventions on climate change, biodiversity, desertification and for the Intergovernmental Panel on Climate Change. (Holden, Roseland, Ferguson, et al., 2008)

Most definitions of sustainability focus on the difference between growth and development. It is pointed out that the difference lies in that growth implies quantitative physical or material increase, while development implies qualitative improvement or at least change (Goodland, 1995). In the books of *The Limits to Growth* (Meadows and Randers, et al., 1972) and *Beyond the Limits* (Meadows and Randers, 1992), Meadows et al concluded that "it is possible to alter these growth trends and establish a condition of ecological and economic stability that is sustainable into the future". Then it is underlined that the sustainable development should integrate social, environmental, and economic sustainability also well known as "the three pillars" of sustainability to make development sustainable. Generally, the three pillars frameworks are called "Social sustainability", "Environmental sustainability" and "Economic sustainability".

The sustainable development is an integrated concept which emphasizes the connection and balance among the three pillars, rather than only conserving the environment or reducing human's activities, which is supported by a lot of people. Sustainability does not only means conserving the natural resources but also means making economic and social achievements such as poverty reduction which is thought to be the most important goal of social sustainable development, even more important than the environmental quality fully addressed (Redclift, 1994). The widely accepted definition of economic sustainability is the maintenance of capital or keeping capital intact, which could be described as "the amount one can consume during a period and still be as well off at the end of the period" (Hicks, 1946). The target of environmental sustainability is to protect the capacity of the human life-support systems-environment, ecology, sources etc.

2.2 The research of green building and assessment

2.2.1 The concept of green building

Now, green building has become a flagship of sustainable development in this century that takes the responsibility for balancing long-term economic, environmental and social health. (Yoon S et al, 2003) In Wikipedia, green building, also known as green construction or sustainable building, is defined as a structure and using process that is environmentally responsible and resource efficient throughout a building's life-cycle: from scheme to design, construction, operation, maintenance, renovation, and demolition which means that close cooperation and participation of the design team, the architects, the developers, the engineers, and the clients are required at all project stages. (Yan Ji and Stellios Plainiotis, 2006) In addition, compared with those traditional buildings, Green building is designed to meet more needs and requirements. It has to protect occupants' health; improve users' productivity; consume less resource; and reduce waste, pollution and environmental degradation." (U.S. Environmental Protection Agency, 2009)

As a matter of fact, the idea of green building has a history of more than 50 years until now. In 1969, Paolo Soleri created the concept of "arcology" which combines the two disciplines of ecology and architecture for the first time. Then in the energy crisis of the 1970's, green building began to practice in reality from the research of theory. At that time, the builders and designers were looking for a way to reduce the reliance on fossil fuels of buildings and hopes to solve the problem brought by the energy crisis, and make the building more environmentally friendly. As a result, a number of new buildings that employed the main concepts of Green Building were constructed in 1970's, such as The Willis Faber and Dumas Headquarters in England, which utilized a grass roof, day-lighted atrium, and mirrored windows; the Gregory Bateson Building in California, which used energy-sensitive photovoltaic (solar cells), under-floor rock store cooling systems, and area macroclimate control devices. Then in 1980's, much researches were commissioned on the energy efficient processes, which resulted in high effective solar panels, efficient prefabricated wall systems, water reclamations systems, modular construction units, direct usage of light through windows, etc. All these techniques were finally implemented in the construction practice with the purpose to decrease the all day-time energy consumption of the buildings. (Web of Top Energy, 2007)

In 1990's, the main concepts of green building were well accepted by the newborn global sustainable movement and then the fast development of the green building gave birth to a series of assessment tools and third-party rating systems all over the world. Generally speaking, all green building assessment tools and rating systems share the essential principles of environmental protection, energy efficiency, water conservation, etc. They often focus on the same building design and lifecycle performance categories of site, water, energy, materials, and indoor environment. (Web of Top Energy, 2007) In order to reduce the complexity, most tools and rating systems use a labelling or certification way to measure the sustainability of buildings assessed. However, the documents, certification procedures, etc. of each tool and system are different because of their various application contexts.

2.2.2 Green building assessment tools and rating systems

Now, the building sector has become the largest energy consumer and source of greenhouse gas emissions around the world. Being green, or sustainable, is one of the most urgent issues coming from internal and external drivers for both nations and enterprises.(Wu and Low, 2010) Obviously, such a status offers an opportunity to make buildings more environmentally friendly and efficient with an integrated approach to reduce the negative impact of building on the environment and occupants. In order to measure how green, or sustainable a building is, a series of green rating systems have been developed, such as Leadership in Energy and Environmental Design (LEED), Building Research Establishment's Environmental Assessment Method (BREEAM), Comprehensive Assessment System for Building Environmental Efficiency (CASBEE), Green Building Tool(GBTool), etc.

All these rating systems provide an effective framework for assessing building environmental performance and integrating the concept of sustainable development with building and construction processes. These systems can be used as a design tool by setting sustainable design priorities and goals, developing appropriate sustainable design strategies; and determining performance measures to guide the sustainable design in the decision-making processes. (Cole R, 2003) The management utility of the tool, such as establishing organizational structure, building commissioning, and so on, during the design, construction and operations also cannot be neglected (Wu and Low, 2010).

Green building design does not only mean a positive impact on public health and the environment, it is also able to reduce the operating costs, enhance building and organizational marketability, increases occupant productivity, and help create a sustainable community (Fowler

KM and Rauch EM, 2006). Generally speaking, green building rating systems focus on the following five categories of building design and lifecycle performance: site, water, energy, materials, and indoor environment. They often consist of a number of prerequisites and credits with specific design and performance criteria, together with a detailed description of the rationale, limitations, and direction for each category.

About the roles the green building assessment tools and rating systems play, it is argued that three kinds of role should be taken into consideration: First, as a means to raise the awareness on the building environmental performance of all the members who participate in the design, construction and operation phases; Second, setting building environmental performance benchmarks to encourage the efforts on enhancing the environmental performance rather than just pass the national standards; And finally, providing a platform for rewarding innovative designs, ideas and techniques.(Cam and Ong, 2005)

Here, another important concept that should be introduced is life cycle assessment (LCA, also known as life-cycle analysis, eco balance, and cradle-to-grave analysis). It is a technique to assess environmental impacts associated with all the stages of a product's life from-cradle-to-grave (i.e., from raw material extraction through materials processing, manufacture, distribution, use, repair and maintenance, and disposal or recycling). (US Environmental Protection Agency, 2010) Now it is widely accepted that the green building assessment and rating system can bring significant benefits which cannot be produced by the standard practices. However, the assessment and rating systems based on building life cycle can produce more significant benefits in long-term, not only for the building owners but also for the occupants for it takes into account all the costs of acquiring, owning, and disposing of a building. (Cole R and Kernan P, 1996) It is used in different areas such as solving existing building problems, limiting environmental impacts, creating healthier places, etc. Generally, such a technique can help the decision-makers make the best choice among different project alternatives which have the same function performance but different initial costs and operating costs.

Now, more and more green building assessment tools emerge to determine the greenness of a building by different ranking systems and generally speaking, there are two types of the existing sustainability assessment tools.

The analysis oriented methods

This kind of methods is developed according to the theory of building's lifecycle which can calculate all of the building's accumulated environmental impacts with some quantitative methods. The most importance of this kind of methods is the database which shows the environmental impacts and their associated weighting of different building materials. The strength of this kind of methods lies in the capability of calculating the real environmental impact of a building in a comprehensive, quantitative way.

The representatives of this kind of methods are Bees (USA), Beat (Denmark), EcoQuantum (Netherlands) and the Athena (Canadian).

The application-oriented methods

The strength of this kind of methods lies in its easy-to-operate characteristic. The application-oriented methods can be defined as a system of assigning point values to a selected number of parameters on a scale ranging between "small" and "large" environmental impact. They often have a check-list based on building lifecycle theory to compare and score all qualitative and quantitative indices of a building's environment impact. Among these methods, the UK Building Research Establishment Environment's Assessment Method (BREEAM), LEED (US)-USGBC, CASBEE (Japan), ITACA (Italy) and GB Tools are well known assessment tools in developed countries. All of them established a comprehensive and efficient framework of assessment criteria for different types of building in their regions. They provided a whole building evaluation rather than an evaluation of some individual design features or techniques and used measurable systems to reveal how much the building incorporate sustainability principles.

2.3 Representatives of international green building assessment systems

2.3.1 Great Britain: BREEAM (Building Research Establishment's Environmental Assessment Method)

BREEAM is a leading and widely accepted assessment method which sets the environmental benchmarks for buildings' performance. The original version of BREEAM (for offices) was developed by BRE, Building Research Establishment, in 1990 for the assessment of buildings for new construction offices. Since then, the method has been continually developed and extended to a family with different members that cover industrial buildings, commercial/retail development, schools, hospitals and domestic buildings (Eco Homes). In addition, those buildings which do not fit into these categories can be assessed using a bespoke form of BREEAM. And in this year, BRE has developed BREEAM for sports and leisure which has been successfully applied to the sports facilities such as those stadiums and buildings for the London 2012 Olympic Games.

The aim of BREEAM

The aim of BREEAM is to reduce the environmental impact of construction, ranging from embodied energy to ecology. Though it is primarily focused on the environmental sustainability, BREEAM also considers some aspects of both social and economic sustainability. The impacts of buildings on the environment are reduced through a given number of voluntary standards which are set higher than building regulations. BREEAM is now well established, accepted and championed by the main stakeholders both in the UK and international.

BREEAM recognizes and encourages the best practice and effort in the design, construction and operation phrases of buildings by awarding a given number of credits according to the level of achievement. It attempts to raise the awareness of the environmental impacts of buildings among the main stakeholders, providing assistance in decision-making when planning, designing, constructing, occupying and managing buildings.

BREEAM offers the potential to reduce operating costs (e. g. energy) and to achieve a better environment for the building users.

The assessment procedure of BREEAM

A national network of independent assessors who are trained and licensed by BRE will make the assessment, working closely with the building design teams or their clients to assess construction with set parameters. Assessments are carried out by assessors at the design stage, after construction, or they can be performed after the building has been in use for two years (operational/management stage assessment).

Then the assessment reports will be sent to BRE for quality assurance. Assessors are building professionals, multi-disciplinary consultancies and building owners and managers. In recent years, the number of assessors as well as the assessment projects is growing rapidly together with the increasing demand for sustainability around the world.

BREEAM rates the performance of a building on a simple six credits scales, "Unclassified" "PASS" "GOOD" "VERY GOOD" "EXCELLENT" and "Outstanding" (Fig 2.1).

BREEAM Communities Rating	% score
UNCLASSIFIED	<25
PASS	≥25
GOOD	≥40
V GOOD	≥55
EXCELLENT	≥70
OUTSTANDING*	≥85

Fig 2.1 The Rating of BREEAM (BREEAM Communities, 2011)

Development procedure of BREEAM

BRE Environmental Assessment Method (BREEAM) is considered as one of the first voluntary measurement rating systems for green building, which were established in the UK by the Building Research Establishment (BRE) in 1990. Since then it has grown in scope and geographically, being exported in various guises across the globe. It stimulated the development of many other assessment methods in different regions such as LEED in North America, Green Star in Australia, HQE in France, ITACA Protocol in Italy, CASBEE in Japan, etc.

Since 1990, the system of BREEAM has experienced a long period of development and update. During the 20 years, the Building Research Establishment (BRE) created different editions of BREEAM based on the practice of assessment and increasing research and theories of green building and sustainable development.

As the demand for green and sustainable buildings is growing rapidly in recent years, BREEAM is widely accepted by the building market. It has become the most important production of environmental stimulation policies in Britain.

Since 1990, BREEAM has completed a lot of valuable work of data collection and analysis. With the development of its different editions, BREEAM has taken a very big sector in the building market of Britain. Now about 25% ~30% of the new office buildings are assessed by BREEAM every year.

2.3.2 United States of America: LEED (Leadership in Energy and Environmental Design)

In 1993, the U.S. Green Building Council (USGBC) was found. After that, all its members had recognized that for the industry of sustainable or green buildings, the most important step is to develop a system to assess the sustainability or greenness of a building. So a committee was established to make research on the existing green building metrics and rating systems in 1994. The careers of the committee members were diverse. They are architects, real estate agents, building owners, lawyers, environmentalists, and industrial representatives.

As part of the interview, the committee made a deep research on two existing green building rating systems of Britain at that time, BREEAM (Building Research Establishment's Environmental Assessment Method) and BEPAC (Building Environment Performance Assessment Criteria). Then three choices faced the committee:

1. Accepting BREEAM ranking system and application in the building market of USA.

2. Revising BREEAM according to the building market of USA.

3. Developing a new green building rating system for the building market of USA.

At the end, the last alternative was chosen by the committee, for they thought that it is necessary to create a green building evaluation tool which is specifically designed for the USA building market. Therefore, in 1998 the first LEED Pilot Project Program was launched. This version was still referred to as the LEED Version 1.0. Then, LEED Green Building Rating System Version 2.0 and following versions were updated after deep modifications and improvements.

Then LEED has improved and matured after several update, the program has began to develop new editions such as LEED for Existing Buildings, LEED for Core & Shell, LEED for New Construction, LEED for Schools, LEED for Retail, LEED for Healthcare, LEED for Homes, and LEED for Commercial Interiors.

Now, LEED has become a big family with different members for specific building typologies, sectors, and project scopes. It has become a system which can address different project developments and processes in the building market. And in 2009, LEED for Neighborhood Development is launched, which is the latest LEED certification system and the third rating system for settlement scale after CASBEE for Urban Development and BREEAM for Communities.

It can be seen from Fig 2.2 that LEED has two kinds of assessment products: for Horizontal Market Products and for Vertical Market Products.

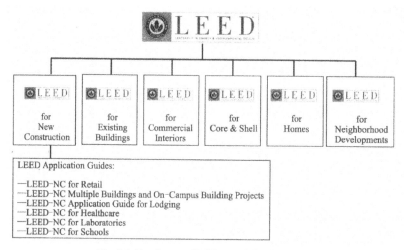

Fig 2.2 Products of LEED (USGBC, 2009)

Horizontal Market Products consist of the 6 main products of LEED, which covers all the phrases of the buildings' whole life cycle and all kinds of buildings. They are the core of the U.S. Green Building Council (USGBC) and take most parts of budget of USGBC. In fact, when USGBC decides to develop a new kind of Horizontal Market Product, it has to consider not only the situations of assessment practice but also the special need of segment market. After that, the final decision which can balance them was made by USGBC. Each Horizontal Market Product should be distinguished from others and provides the same strict standard as the other members of the LEED family. They are developed by their own special committees, in cooperation with the Technical Advisory Groups in order to make sure of the product's consistency. And before published each Horizontal Market Product should be approved by the LEED Steering Committee who vote according to the test assessment projects.

Vertical Market Products are developed to satisfy the market requirements of different kinds of buildings in a same assessment category. In such condition, LEED Application Guides will provide special explanations and detailed information on how to deal with these technological links and how to use the measure promoted by LEED. And finally, all the Vertical Market Products should be approved by the LEED Steering Committee before published.

After the development of more than ten years, LEED has become one of the most important and useful policy tools now. For the local governments of USA, investing green buildings and produce policies stimulating the sustainability development will improve their image in the electors. Therefore the certification of LEED was promoted by them and many benefit tax policies associated.

2.3.3 Japan: CASBEE (Comprehensive Assessment System for Building Environmental Efficiency)

CASBEE is an assessment tool which is developed to assess the building environment performance. It can be used according to the architecture design process, starting from the pre-design stage and then through design and finally post design stages.

In Japan, a joint industrial/government/academic project was initiated with the support of the Housing Bureau, Ministry of Land, Infrastructure, Transport and Tourism (MLIT), in April 2001, which led to the establishment of a new organization, the Japan Green Build Council (JaGBC) / Japan Sustainable Building Consortium (JSBC), with its secretariat administered by the Institute for Building Environment and Energy Conservation (IBEC). JaGBC, JSBC and subcommittees are together working on R&D of the Comprehensive Assessment System for Building Environmental Efficiency (CASBEE).

CASBEE tools are established on the policies as following:

1. High building environmental performance should be rewarded with the purpose to encourage the designers and others to improve and enhance the building environmental performance.

2. The tool should be simple enough.

3. The tool should be able to implement for different building types and in different processes of a development.

4. The tool should consider the local context and problems of Japan.

During the history of ten years, CASBEE has developed and updated different editions and versions of tools corresponding to different stages of building lifecycle. They are shown as following:

CASBEE for Office, which was completed in 2002

CASBEE for New Construction in July 2003

CASBEE for Existing Building in July 2004

CASBEE for Renovation in July 2005.

Each of them is used for an individual purpose and target user group and designed to accommodate a wide range of building types.

CASBEE comprises the four basic tools, tailored to the building lifecycle, and expanded tools for specific purposes. These are called collectively as the "CASBEE Family," for housing scale, building scale and urban scale.

According to CASBEE, the concept of closed ecosystems is essential for determining environmental capacities in assessment practices. Different from other assessment systems, CASBEE is unique for its innovative concept: it assesses a building or development from the two viewpoints of environment quality and performance (Q=quality) and environmental load on the external environmental (L=Load) when evaluating the environmental performance of the building. CASBEE applies the concept of eco-efficiency as BEE (Building Environmental Efficiency) to measure the performance level of assessment project. The scores for Q and L are assigned separately and finally an assessment of BEE as an indicator is given based on the results for Q and L. L is first evaluated as LR (environmental load reduction of the building). That approach is employed because "higher marks for improving load reduction" is much easier to "higher marks for load reduction" as a system, just as "improvements in quality and performance earn higher marks." BEE is calculated from Q and L as shown following:

$$BEE = \frac{Q}{L} = \frac{25 \times (SQ - 1)}{25 \times (5 - SLR)}$$

SQ presents the scoring for Q and SLR presents the scoring for LR. The scoring for Q and LR are both divided into three categories, and each category is determined by several items.

The BEE value assessment result is expressed as the gradient of the straight line passing through the origin (0,0). The higher the Q value and the lower the L value, the steeper the gradient and the more sustainable the building is. Using this approach, it becomes possible to graphically present the results of building environmental assessments using areas bounded by these gradients (Eco-labeling). The assessment results for buildings can be labeled on a diagram as class C (poor), class B−, class B+, class A, and class S (excellent), in order to increase BEE value.

The Q and LR consist six sub-categories which are:

- Indoor environment (Q−1)
- Quality of Service (Q−2)
- Outdoor Environment on Site (Q3)
- Energy (LR−1)
- Resource & Materials (LR−2)
- Off-site Environment (LR−3)

Scores are given based on the scoring criteria for each assessment sub-categories. These criteria applied to assessments are determined by the consideration of the level of technical and social standards at the time of assessment. A five-level scoring system is used, and the score of level 3 indicates an "average" performance.

Each assessment sub-categories, such as Q-1, Q-2 and Q-3, is weighted so that all the weighting coefficients within the assessment category Q sum up to 1.0. The scores for each assessment sub-categories are multiplied by the weighting coefficient, and aggregated into SQ; total scores for Q or LR; total scores for LR respectively.

2.3.4 Italy: ITACA Protocol

The ITACA Protocol is an assessment tool of the building performance on the energetic-environmental sustainability. It is approved in 2004 by the Inter-Regional Work Group which formulates a series of rules and standards shared on a National level for the definition of project with bio-construction features.

Now most Regions and Autonomous Provinces in Italy complied with the ITACA Protocol. In addition, they have used it as an instrument and reference to promote sustainable development and to stimulate the economic incentives and encouragement to those who make efforts on the green buildings.

The ITACA Protocol was developed in the field of the Green Building Challenge process, which was developed on the international level by the United Nations Environment Program Sustainable Building & Construction Initiative (UNEP-SBI).

The new edition of ITACA Protocol was adopted by the Working Group interregional "Sustainable Building" on February 25, 2009, according to the complete structure of the protocol of 2009 which was already adopted by the Working Group on 16 December 2008.

ITACA Protocol was used to assess the environmental sustainability level of a residential building by measuring its performance, according to 18 categories, and then counting in 5 evaluation aspects:

1. Quality of the site

    1.1 Condition of the site

    1.2 Accessibility of serves

2. Resources consumption

    2.1 Non-renewable primary energy required during the life cycle

    2.2 Energy from renewable sources

2.3 Eco-friendly materials

2.4 Drinking Water

3. Environmental loads

3.1 Emissions of CO_2 equivalent

3.2 Wastewater

3.3 Impact on the surrounding environment

4. Indoor environmental quality

4.1 Venting

4.2 Welfare thermo hygrometric

4.3 Healthy sight

4.4 Welfare acoustic

4.5 Electromagnetic Pollution

5. Service quality

5.1 Controllability of plants

5.2 Maintenance of performance in operational phase

5.3 Areas of the building

5.4 Home Automation

For each criterion, the building gains the credits that vary from −1 to +5, which are assigned by comparing the assessment indicator with the values of the performance scale previously defined. And zero represents the average performance or the standard level according to the laws and the regulations enforced or.

2.4 Research of sustainable residential area and assessment

2.4.1 Concept of sustainable residential area

A residential area is a settlement in which land use of housing or residence predominates, as opposed to industrial and commercial areas. In residential area, public supporting facilities (like shops, schools, entertainments, etc) are supplied, and commercial services and work opportunities usually emerge to meet the residents, demand. (Wikipedia, 2008) General speaking, a community, a neighborhood or a residential district (which has a special definition in the national standards of China) can be all considered as a kind of residential area, and their names are better accepted by the public than the residential area.

Until now, there has been still no definition of sustainable residential area. But as a kind of residential area, sustainable communities are better defined. The Government of the United Kingdom defines a sustainable community in its 2003 Sustainable Communities Plan: "Sustainable communities are places where people want to live and work, now and in the future. They meet the diverse needs of existing and future residents, are sensitive to their environment, and contribute to a high quality of life. They are safe and inclusive, well planned, built and run, and offer equality of opportunity and good services for all." (Maliene V, et al, 2009)

Wikipedia defines the sustainable communities as those which are planned, built, or modified to promote sustainable living style. This may include sustainability aspects relating to reproduction, water, transportation, energy, and waste and materials. (California Sustainability Alliance, 2010) Generally speaking, they pay more attention on the environmental sustainability and economic sustainability. The sustainable urban infrastructure and/or sustainable municipal infrastructure are also focused on by sustainable communities. (Wikipedia, 2008)

Though, the definition and concepts are still in discussion, the attempts and practices of sustainable residential area are emerging consecutively. Many good projects of sustainable residential area are constructed all over the world and promoting the urban sustainable development to a great extent.

2.4.2 Attempts of sustainable residential area practice

Beddington Zero Energy Development (BedZED), London, UK

The Beddington Zero Energy Development (BedZED) is the UK's largest mixed-use sustainable community. It was located in the London Borough of Sutton, on the suburban fringe of the city, and completed and occupied in 2002.

The project was designed by the architect Bill Dunster with a purpose to promote a more sustainable lifestyle. It was led by the Peabody Trust together with Bill Dunster Architects, Ellis & Moore Consulting Engineers, BioRegional, Arup and the cost consultants Gardiner and Theobald.

The community comprises 99 homes and 1,405 square meters of work space, which were built from 2000 to 2002. The 99 homes comprise 50 per cent housing for sale, 25 per cent key worker shared ownership and 25 per cent social housing for rent (Chance T, 2009).

BedZED was designed to be a thriving community in which ordinary people could enjoy a high quality of life while living within their fair share of the earth's resources (BioRegional Development Group, 2007). Ecological and carbon footprinting were used to set benchmarks for "sustainable lifestyles". The average ecological footprint in Sutton is 5.32 global hectares per person compared to a sustainable limit of 1.7 global hectares per person. Average carbon dioxide emissions in Sutton are 11.17 tonnes per person per year compared to a sustainable limit of between one and 1.5 tonnes. (Wiedmann et al, 2006) (i.e. the global average that, if achieved, should avoid dangerous levels of climate change).

BedZED not only tackled the carbon emissions from domestic and office energy use but also talked from the carbon emissions arising from the building materials producing, transport, food and waste. In addition, the strategy also covered the issues of water, quality of life and strengthening the local economy. This approach makes BedZED support the best sustainable living style in the world. In 2003, the project was shortlisted for the Stirling Prize. (Wikipedia, 2012)Then after the occupants moved into the homes and offices of the community, BioRegional has begun to conduct several monitoring and evaluation exercises in order to make a deep research of this community project. It was proved that though when the people move to BedZED they have common lifestyles, their behavior and lifestyles had been changed significantly after living there for years.

BedZED enables a reduction in greenhouse gas emissions of the community according to the UK's obligations. Many techniques of green building were adopted in the building design of BedZED . In addition, the designer solved the problems of heating and water usage, provided more opportunities to the residences to make sustainable choices and created more facilities and groups to enhance the life quality of the residence with less environmental impact.

Now, the community has become the home to approximately 220 residents and 100 office workers. The original idea of the mix-used community design was that such a composition of residence and work space will create a new lifestyle that allows people to work and live together and reduce the personal transportation to a big extent. However, such units didn't sell very well and were finally sold as private residences.

Good space making and community management is the most popular aspect of BedZED. The designer planned various public spaces around BedZED, which consisted of the pedestrian streets, a small square for young children, a sports playground and a community centre. With such facilities, the community of BedZED is much stronger than other communities. In BedZED, the residents know 20 neighbors on average. The site plan of BedZED is shown in Fig 2.3.

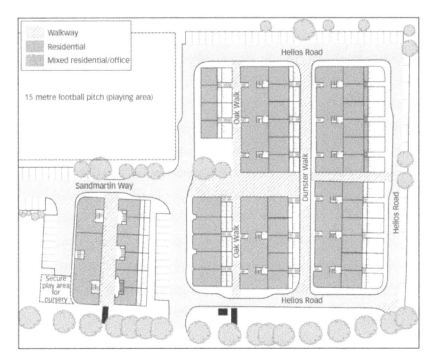

Fig 2.3 Site plan of BedZED (BioRegional, 2003)

Vauban, Freiburg, Germany

In German, the word Rieselfeld means sewage farm. In fact, this area had been used to treat sewage for more than 100 years since 1891 after the leaders of Freiburg bought the area from the University of Freiburg. Vauban is located 4 km to the south of the town center in Freiburg, which is a new neighborhood planned for 5,000 inhabitants and 600 jobs when the project finished (www.vauban.de).

The planning for Vauban started in 1993, with a purpose to develop this area as a sustainable model district. The construction began in the mid-1990s and completed in 2006, after a three-phase development. The final scale of Vauban sustainable model district is 38 hectares. Different from other project, the Vauban project is developed in a cooperative and participatory way. The citizen's association Forum Vauban e.V., which has NGO-status, was applied to organize the participation process and was recognized as its legal body by the City of Freiburg in 1995 (www.vauban.de). The photos of Vauban before and after the development are shown in Fig 2.4.

The sustainable district of Vauban was established by the new sustainable urban life styles which were created by the public space plan and building design. In Vauban, all the buildings are designed

Fig 2.4 Vauban before and after the development (Wang, 2008)

according to a strict standard on energy saving. 100 residences are designed to the Passivhaus ultra-low energy building standard. And other buildings are heated by a combined heat and power station burning wood chips, while many of the buildings have solar collectors or photovoltaic cells (Passive House Institute, 2006). Perhaps the best example of sustainable building is the Solar Settlement in Vauban, a 59 Plus Energy home housing community(Fig 2.5). It is the first housing community worldwide in which all the homes produce a positive energy balance. The solar energy surplus is then sold back into the city's grid for a profit on every home.

Fig 2.5 The Solar Settlement in Freiburg (http://www.rolfdisch.de/)

Transport is another feature of Vauban project. In Vauban, foot or bicycle is the primary transportation. While, in order to connect with the city center of Freiburg a tram line (Fig 2.6) was constructed and the development was planned along with the line to ensure all the residence can be covered within a walking distance radius.

With the convenient public transportation system, the level of car ownership has fallen over time. An earlier survey showed over 50% of households owned a car; of those who were living without a car, 81% had previously owned one and 57% gave up their cars on or

Fig 2.6 The tram line in Vauban (Wang, 2008)

immediately after moving to Vauban (Nobis. C, 2003).

Sino-Singapore Tianjin Eco-city, Tianjin (Tientsin), China

The Sino-Singapore Tianjin Eco-city is the result of a collaborative agreement between the governments of China and Singapore in response to global climate change. The framework agreement was signed by Chinese Prime Minister Wen Jiabao and Lee Hsien Loong, the Prime Minister of Singapore in 2007. As an innovative demonstration project aiming at sustainable development, the Sino-Singapore Tianjin Eco-city's vision is to be a thriving city which is socially harmonious, environmentally-friendly and resource-efficient, which will set a model for all Chinese cities.

The Eco-city locates on non-arable land. Prior to development, the site of the Eco-city was one-third saltpan, one-third deserted beach, and one-third water, including a 270 hm^2 wastewater pond. The total planning area is about 31 km^2. When fully developed in 10 to 15 years' time, the eco-city will be home to about 350,000 residents.

In the planning of the Tianjin Eco-city, one of the main guiding principles was to adopt a approach towards creating a livable, efficient and compact city, which would be developed into an ecologically sound and environmentally sustainable manner, which created some distinguishing features of the Sino-Singapore Tianjin Eco-city.

The natural ecological wetlands will be conserved and the historical water bodies will be rehabilitated. The main center with multi-functions, including commercial, cultural and recreational uses of the Eco-city will be located on the southern bank of an old river. In addition, a comprehensive green transport network will be developed in the Eco-city, which encouraged the residents to reduce the automobile reliance. A light rail transit system will serve as the main mode of transport through the Eco-city. Green trips, which include public transportation, cycling and walking, will also be promoted in the Eco-city. The target is for at least 90 per cent of the trips within the Eco-city to be via walking, cycling or use of public transport. The use of clean fuel and renewable energy such as solar energy and geothermal energy will be explored in the Eco-city also.

Until now, the initial area of the Eco-city has been completed (Fig 2.7).

Fig 2.7 The initial area of Eco-city
(http://www.tianjinecocity.gov.sg)

2.5 Representatives of existing sustainable residential area assessment tools

2.5.1 From green building assessment to residential area assessment

After the green assessment of individual building were well accepted by the markets, most representative green building assessment systems began to explore a special tool for the sustainability assessment of big scale or area development. Until now, BREEAM Communities (2009), LEED for Neighborhood Development (2009) and CASBEE for Urban Development (2007) have been the three existing sustainability assessment tools for residential area development. The basic concepts and ideas of sustainable communities mentioned before are well accepted by these systems. These new established tools can measure the sustainability of an area development and with a rating system to determine how sustainable the project assessed is.

Now, the environmental assessments are commonly adopted in large-scale development projects. Different from the environment assessments, these sustainable residential area assessment tools have different concepts and roles in the big scale developments. They address the importance of various actions against environmental, economic and social problems and make a comprehensive assessment of the merits and demerits on the environmental, economic and social sustainability of a development.

Here, the three famous existing assessment tools for residential area development, BREEAM Communities, LEED for Neighborhood Development and CASBEE for Urban Development, will be introduced here briefly and a further comparison and analysis of them will be made in later chapters.

2.5.2 BREEAM Communities

As mentioned before, BREEAM Communities is an important member of BREEAM family. BREEAM Communities employs the established BREEAM methodology and is an independent, third party assessment and certification standard of a residential area scale development.

BREEAM Communities was developed to help the design team, developers and planners to improve, measure and independently certify the sustainability of developments at the neighborhood scale or larger and first version was launched on 2009.

The aims of BREEAM Communities

As the manual written, the aims of BREEAM Communities are (BRE, 2009):

• To mitigate the overall environmental impacts of area development projects

• To enable the development projects to be recognized according to their environmental, social and economic benefits to the local community

• To provide a credible and holistic environmental, social and economic sustainability label for development projects in the built environment

• To stimulate demand for sustainable development (and sustainable communities) within the built environment

• To ensure the delivery of sustainable communities within the built environment

The size definition in BREEAM Communities

Different from individual building, size, scale or boundary is an important feature for the assessment of residential area development. In BREEAM Communities, the scale definition is simpler and more accurate than the other two existing tools. The community is rated into 4 classes according to the building numbers. It is shown in Table 2.1 and for the different scales of area development there are compliant assessment items which can solve the criteria applicability problems faced in the assessment practices.

Size	Number of units
Small	up to 10 units
Medium	between 11 and 500 units
Large	up to 5,999 units
Bespoke	6,000 or greater

Table 2.1 Size of Developments (BRE, 2009)

Assessment stages of BREEAM Communities

BREEAM Communities has different assessing stages from those building scale assessment members in BREEAM family. In BREEAM Communities, assessing the environmental impacts arose as a result of a site wide development at the following stages as shown in Table 2.2.

Assessment items of BREEAM Communities

In BREEAM Communities, Credits are awarded in eight categories of sustainability according to their performance against the sustainability objectives and planning policy requirements. These credits are then added together to produce a single overall score on a scale of Pass, Good, Very Good, Excellent and Outstanding. And it has eight categories.

Name of stages	Time	Description
Registration of BREEAM Communities	Create Compliant Assessment Framework	The registration of a "compliant assessment framework" is the first step and a mandatory requirement if a development wants to achieve certification against the BREEAM Communities standard.
Interim BREEAM Communities Certificate (Optional)	Completed at the Outline Planning Stage (OPS)	The preliminary planning stage assessment and subsequent "Interim" Certification measures the commitments outlined at the preliminary planning stage against the key sustainability objectives and planning policies, within the applicable local planning system.
Final BREEAM Communities Certificate (Mandatory)	Completed at the Detailed Planning Stage (DPS)	The final planning stage assessment and subsequent "Final" Certification measures the detailed commitments outlined within the final planning stage application against the key sustainability objectives and planning policy requirements within the applicable local planning system.

Table 2.2 Assessment stages of BREEAM Communities (BRE. 2009)

2.5.3 LEED for Neighborhood Development

Different from other members of LEED family, the LEED for Neighborhood Development Rating System is developed by the The U.S. Green Building Council (USGBC), the Congress for the New Urbanism (CNU), and the Natural Resources Defense Council (NRDC), with a purpose of establishing a national leadership standard with which to assess and reward those neighborhood development practices which have good environmental performance, within the framework of the LEED Green Building Rating System. It integrates the principles of smart growth, new urbanism and green building into a rating system for neighborhood planning and development.

The aims of LEED for Neighborhood Development

Unlike other LEED rating systems on building scale, which focus primarily on green building practices and offer only a few credits for site selection and design, LEED for Neighborhood Development focuses on the site selection, design, and construction elements that bring buildings and infrastructure together into a neighborhood and relate the neighborhood to its landscape as well as its local and regional context. Because the establishment process was under the guidance of the combined concepts of the three participant organizations, smart growth, New Urbanism, and green infrastructure and building, it is not only a rating label tool but also an effective guideline for decision, design and construction of neighborhood development.

The definition of neighborhood development in LEED for ND

As one of the three partners, the Congress for the New Urbanism (CNU) has a great effect in creating the LEED for Neighborhood Development. Some important concepts and ideas of New Urbanism are introduced in, and neighborhood is one of them.

A neighborhood can be considered the planning unit of a town. The charter of the Congress for the New Urbanism characterizes this unit as "compact, pedestrian-friendly, and mixed-use" (CNU, 1996). The neighborhood can be considered as a village, but combined with other neighborhoods it becomes a town or a city. Similarly, several neighborhoods with their centers at transit stops can constitute a transit corridor. The neighborhood, as laid out in LEED-ND, is in contrast to sprawl development patterns, which create pod like clusters that are disconnected from surrounding areas. Instead, traditional neighborhoods meet all those same needs—for housing, employment, shopping, civic functions, and more—but in formats that are compact, complete, and connected, and ultimately more sustainable and diverse. (CNU, 1996)

Generally speaking, the size of a neighborhood in LEED for Neighborhood Development is defined according to the comfortable distance for walking from the center to the edge of the neighborhood. After a review and analysis of related theory, half a square mile contains 320 acres (1.29 km^2 or 129 hm^2) is defined as the size should serve as guidance for the upper limit of a LEED-ND project by the core committee. And in the assessment practice, if the project is larger than that size, it is necessary to divide the project into parts small than that size to assess.

Assessment stages

In LEED for Neighborhood Development, because projects often have significantly longer constructions periods than single buildings, the standard LEED certification process has been modified. The certification is divided into a three-stage process.

Assessment categories of LEED for Neighborhood

Unlike the other LEED rating systems which have five environmental categories, LEED for Neighborhood Development has three:

Smart Location and Linkage

Neighborhood Pattern and Design

Green Infrastructure and Buildings

In addition, LEED-ND has another two categories, Innovation and Design Process which addresses sustainable design and construction issues and measures not covered under the three categories and Regional Priority Credit which acknowledges the importance of local conditions in determining best environmental design and construction practices as well as social and health practices. (LEED-ND, 2009)

2.5.4 CASBEE for Urban Development

After succeeding in developing CASBEE for building scale, the Japan Sustainable Building Consortium (JSBC, with the Institute for Building Environment and Energy Conservation as secretariat) began the research to adapt CASBEE for urban scale assessment. In July 2006, the new tool of CASBEE for Urban Development which was developed to assess the environmental performance of groups of buildings on urban scale launched.

The aims of CASBEE for Urban Development

In CASBEE family, CASBEE for Urban Development acts as a general rule which controls the development of whole groups of buildings rather than individual ones. As mentioned in the technical manual of CASBEE for Urban Development, its purpose is that "When a project is planned and implemented that comprises multiple buildings and other elements on a single, large-scale site under a unified design concept, assessment can go beyond the environmental design of each building, to identify new or expanded environmental measures, and their effects, that are made possible by the building group, and thereby contribute to the comprehensive improvement of environmental performance in urban renewal."

CASBEE for Urban Development is one of the expanded CASBEE tools, developed with reference to the Q3 (Outdoor Environment on Site) and LR3 (Off-site Environment) assessment items of CASBEE for New Construction. However, it is an independent system which places concentration on the outdoor space and phenomena that can accompany the conglomeration of buildings. The interior of building is out of its assessment scope, which makes it work independently without any collision with other CASBEE tools for building scale. At the same time, a tool of CASBEE for an Urban Area + Building is also developed to assess the urban area as a whole and the buildings. The CASBEE for an Urban Area + Buildings is not an independent tools, it works together with building scale and urban scale.

The definition of boundary in CASBEE for Urban Development

Different from other two assessment systems, CASBEE has a very special concept— the hypothetical boundary—both in the individual building and the urban development edition. When setting the hypothetical boundary as the designated area, the boundary should be based on the extent (or boundary lines) of the site. When working on the building scale, the designated area is naturally the building site for the project (the site boundary). On the urban scale, the approach described below for the designated area directly reflects the "unified design concept," and is objective and clearly comprehensible.

1. In CASBEE for Urban Development, setting the area for the project to be assessed should, in principle, follow the plan area or project area etc. as stipulated by various ordinances, systems and methods etc. applied to the planning and development of the project concerned.

2. The systems and methods that could be applied here include urban redevelopment projects, land readjustment projects, urban renaissance special area, various types of district plans, comprehensive design of the area development projects, and the building design system.

3. As an exception, however, where deemed appropriate for comprehensive environmental performance assessment on the urban scale, adjacent areas outside the above areas can be taken into the scope of the assessment designation, or conversely, part of the area can be excluded from the evaluated area. If an exception is applied under this clause, the assessor must clearly state the reason for doing so.

Assessment stages

In CASBEE for Urban Development, assessments generally require knowledge in many expert fields, and some of the assessment items demand a certain level of planning precision. In order to solve the problem, two versions are set for the different assessment stages as shown in Table 2.3.

Name of stages	Description
Brief version	For the initial planning stage
Standard version	For the mature planning stage

Table 2.3 Assessment stages of CASBEE for Urban Development (CASBEE for UD. 2009)

Assessment items of CASBEE for Urban Development

Different for BREEAM and LEED, all the assessment items of CASBEE are divided into two parts: Q (quality) and L (Loadings). For the CASBEE for Urban Development, the Q (Q_{UD}) refers to the environmental quality within that boundary, while the L (L_{UD}) refers to the environmental load beyond the boundary. Because the assessment of buildings is included in the building scale CASBEE tools, it is not covered by CASBEE for Urban Development. Both the Q_{UD} and the L_{UD} have three categories which are:

$Q_{UD}1$ Natural Environment (microclimates and ecosystems)

$Q_{UD}2$ Service functions for the designated area

$Q_{UD}3$ Contribution to the local community (history, culture, scenery and revitalization)

$LR_{UD}1$ Environmental impact on microclimates, façade and landscape

$LR_{UD}2$ Social infrastructure

$LR_{UD}3$ Management of the local environment

After the assessment, the score of these six categories is compounded with the formula below to generate the Building Environmental Efficiency in urban development (BEE_{UD}).

$$BEE_{UD} = \frac{Q_{UD}}{L_{UD}} = \frac{25 \times (SQ_{UD}-1)}{25 \times (5-SLR_{UD})}$$

In the same way as building scale CASBEE tools, CASBEE for Urban Development uses LR_{UD} (Load Reduction) assessed first to calculate L_{UD}. The final ranks depend on the final results of BEE_{UD} and Q_{UD} according to the rules as shown in Table 2.4.

Ranks	Assessment	BEE value, etc.	Expressions
S	Excellent	BEE=3.0 or more, Q=50 or more	★★★★★
A	Very Good	BEE=1.5~3.0	★★★★
B+	Good	BEE=1.0~1.5	★★★
B-	Fairly Poor	BEE=0.5~1.0	★★
C	Poor	BEE=less than 0.5	★

Table 2.4 Correspondence between ranks based on BEE values and assessments (CASBEE for UD. 2009)

Chapter Three: Assessment Scale and Platform Study of Sustainable Residential District Assessment in China

第3章 我国大型城镇住区可持续性评估的尺度与平台研究

我国在可持续发展与绿色建筑领域的起步较晚，相较于当时发达国家的绿色建筑及可持续技术发展水平，仍存在一定差距。为了弥补这一差距，住建部陆续制定并颁布了多项重要文件，诸如《国家示范工程建设技术要点》《绿色住宅指南及技术要点》等，为我国的绿色建筑发展提供了明确的指导方向。随着绿色建筑理论的不断深化与实践的推进，我国绿色建筑的数量与质量均得到了显著提升。

同时，我国在绿色建筑与可持续发展领域的认知和技术层面尚存在若干挑战，科学有效的可持续发展设计机制的缺失，将会对城市的长远发展产生深远影响。因此，为了推动城市的可持续发展并有效控制建设开发活动，我国亟须建立一套系统、高效的机制，以指导并规范城市可持续发展设计的实施。这一机制的建立对于促进城市建设的绿色转型、提升生态环境质量、保障经济社会的可持续发展具有重要意义。

中国在绿色建筑评估领域已建立了3个重要的评估系统。首先是21世纪初由多家权威单位，包括中国工业联合会、清华大学、建设部技术发展促进中心、中国建筑科学研究院以及哈尔滨工业大学等，共同编制的《中国生态住宅技术评价手册》，这是我国首个针对生态住宅建筑的评估体系。与此同时，清华大学、中国建筑科学研究院、北京建筑设计研究院以及中国建材科学院等单位联合启动了北京奥运会绿色建筑评价体系（GOBAS）的研发工作。《绿色建筑评价标准》的颁布和"三星体系"的推出，也标志着我国绿色建筑发展的自愿性评级机制的出现。

我国在绿色建筑评估工具研究方面取得了显著进展，但仍面临诸多挑战。首要问题在于，由于国内对生产过程中不同材料能耗评估的基础研究和数据支持相对匮乏，评估工具中的多项指标难以量化，这导致现有的评估工具在实际应用中存在一定的局限性。此外，大型住区开发对可持续性评估工具的需求日益迫切，然而目前我国在此领域的研究尚属空白。考虑到我国的经济、政治和社会背景，评估工作应寻求一个恰当的平台和规模，以确保评估工具能够被市场广泛接受并发挥其应有的作用。

3.1 Sustainable development and green building in China

3.1.1 Status quo of sustainable development and green building in China

Since 1990's, the concept of green building and sustainable development has been introduced into China. In 1994 China's Agenda 21 which clarifies China's sustainable development strategies and policies was published. It has been formulated to correspond with Agenda 21 and reflect the Chinese situation. In 1996 *China Report on the Development of Human Settlement* was published. It provided more measures and requirements to improve the residential environment.

Compared with the developed countries, China is still poor in the development of green building and sustainable technologies especially in the green building design, natural ventilation, renewable energy, green building materials, resource recycling, planting technology and etc.

However, since 2001 Ministry of Housing and Urban-Rural Development of the People's Republic of China (MOHURD) has formulated several documents like the main points of national demonstration project construction techniques, guidelines and technical points of green residential buildings and etc. Moreover, in 2006 the Evaluation Standard for Green Building of China was published.

With the development of green building theory, more and more green buildings are constructed in China especially in some big cities like Beijing and Shanghai. According to the national standard of China, the area of annual new construction energy saving building had risen from 10 million m^2 in 1996 to 50 million m^2. Beside that in the sector of renewable energy, China also made a progress. Solar water heater is very common in China. Since now the total panel area of solar water heater has reached 26 million m^2 with a 25% annual increase rate. And some renewable energy technologies like geothermal heat pump is also in fast development. (MOHURD, 2005)

Problems in acknowledge of green building and sustainable development

Now there has been 10 years since the concept of green building and sustainable development was introduced into China for the first time. During the past 10 years, this concept has been widely accepted by Chinese people who believe green building and sustainable development represent the need of world's future. But there are many differences of opinion on how to definite the "Green building" which leads to a misunderstanding in this concept for the whole society in China.

Experts' opinion

Some experts define "Green Building" as a set of idea targets which maybe cannot implement now but means the direction of development. Another opinion considers "Green Building" as a dynamic definition which means that for a special building Green is a temperate concept in each phrase of building's life cycle.

Customers' opinion

Recent years, the concept of green building and sustainability has been accepted by more and more people. Most of them are not very familiar in this area and not real environmentalist but they would like to pay more for the technology of green buildings.

In 2005, Tsinghua University has made a questionnaire survey on the consumer's views on the green buildings. The results showed that the public attention on the pan-green products such as energy saving electronic applications, water saving applications, renewable applications, etc. has risen a lot. Many of them consider the eco value as one of the most important factors when they decide to buy a house and 1%~10% increase of the price due to the ecology and environment can be accepted by them.

Different opinions of Green Building between governments and estate developers

The development of green building in China has to face a lot of difficulties such as the big difference in economy, environment between regions and cities. Now in China the central and municipal governments play a more powerful role than other countries' governments on developing green buildings. On the other hand, in China the estate developers have a high profit market especially for the top market. Therefore, the governments' opinions on green building are much different from the estate developers'.

Most government offices and institutes encourage the low-price green building technologies because they want to improve the environmental and energetic performance of the whole country. However, for the estate developers they generally tend to choose some high technologies which could raise the quality of their projects and catch the customers' eyes to help them get a higher profit ratio.

Problems in implementation of green building and sustainable technologies

Base on "the investigation of status quo and trend of Chinese green building", five problems in implementation of green building and sustainable technologies were concluded as following (Top Energy, 2007):

1. Some technologies are advanced in the world while more others fall behind.
2. System of technological standards is still in establishment process.
3. Many gaps exist in the basis technology research.
4. Many foreign technologies and methods should be localized.

5. The mechanism of sustainable design needs to be established.

All these five problems are obstructions for green building and sustainable development in China. However, among the five problems the absence of sustainable design mechanism has the greatest influence due to the fast urbanization and huge amount of new constructions. As a result, now it is very urgent and necessary to establish an efficient mechanism to guide the sustainable design and control the construction in China.

3.1.2 Status quo of green building assessment in China

Until now, there have been three green building assessment systems in China. Here an introduction of them is shown as following:

The Technical Assessment Handbook for Ecological Residence of China

The Technical Assessment Handbook for Ecological Residence of China was developed by the China Federation of Industry Real Estate Chamber of Commerce, Tsinghua University, Ministry of Construction Technology Development Promotion Center, China Academy of Building Research, Harbin Institute of Technology and etc in 2001. It is the first assessment system of ecological residential buildings.

The assessment system has a framework of four indicators levels and the top level consists of five items: Environmental Design, Energy and Environment, Indoor Environmental Quality, Water and Material and Resources and the bottom level is special measures. This kind of framework has a good flexibility, for it is very easy to add new indicators to the system. Because in China the basis data of assessment like the energy consumption of different materials producing is still unavailable, many of these indicators are qualitative.

Assessment system for Green Building of Beijing Olympic

Since 1970's, International Olympic Committee had began to request on the environmental protection for every Olympic Games host city and made it as a stable policy. In order to hold a eco Olympic Games 2008, In 2002 Tsinghua University, China Academy of Building Research, Beijing Institute of Architectural Design and Research Institute, China Building Materials Academy began to develop the Assessment system for Green Building of Beijing Olympic (GOBAS). And in 2003, the system (GOBAS) was established.

In the concept of assessment method, GOBAS is very similar with CASBEE which applies the concept of eco-efficiency as BEE (Building Environmental Efficiency) to measure the performance level of assessment project and assesses a building or development from the two viewpoints of Q (Quality) which means the environmental quality and service quality, and L (Load) which means the load of energy, resources and environment. In CASBEE, the scores for Q and L are assigned separately and finally give an overall assessment result of BEE as an indicator based on the results for Q and L.

GOBAS is composed of four parts: Green Olympic Building Assessment Program (about the assessment items and requirements), Green Olympic Building Rating Guide (about the assessment method), Green Olympic Building Rating Manual (the special introduction of assessment theory) and assessment software.

The assessment procedure of GOBAS is defined as four phrases:

1. Site Planning
2. Building Designing
3. Construction
4. Project Acceptance and Operating

According to the different characters and requirements GOBAS established a set of assessment indicators including four quantitative indicators of materials: energy consumption, resources consumption, environmental impact and localization. Besides that, GOBAS established a framework of indicators which can assess the building energy saving level, the air pollution, the carbon emissions, etc. These indicators are very useful for the future research of Green Building Assessment of China. And GOBAS had been used in 12 projects of Beijing 2008 Olympic Games.

Standard by People's Republic of China Evaluation Standard for Green Buildings

The Three Star System (Standard by People's Republic of China Evaluation Standard for Green Buildings) published by the Ministry of Construction's Green Building Evaluation Standard is first attempt to create a national green building standard in China. As the introduction of this rating system notes, the purpose of the Evaluation Standard for Green Buildings is to create a voluntary rating system that will encourage green development in China.

Now China is in the phase of rapid economic development and urbanization. In recent years China ranks first in terms of annual building volume, with significantly growing consumption of resources year by year.

Therefore, it is necessary to implement the concept of sustainable development in process of the fast urbanization and building development. Green building is one of the most important issues of the sustainable development. The purpose of formulating this Evaluation Standard for Green Buildings is to regulate assessment practice on green buildings and promote the development of green buildings. This credit-based evaluation system, introduced in 2006, helps the building developers to choose the best and efficient measure to achieve the target of green building they want to pursue.

The evaluation system has two different versions: one for residential buildings and one for public (i.e. large commercial) buildings. As the rating system described, considering the current construction market in China, this standard will mainly evaluate residential buildings that are huge in quantities and public buildings that consume much energy and resources such as office buildings, mall buildings and hotel buildings. For evaluation on other buildings, this standard can serve as a guideline or reference.

The evaluation standard rates buildings with a variety of prerequisites (called "control items" in the Chinese system) and credits (called "general items" in the Chinese system) in six general categories:

1. Land savings and outdoor environment
2. Energy savings
3. Water savings
4. Materials savings
5. Indoor environmental quality
6. Operations and management

Then the seventh category which is called "Preference items" contains measures that are both up-to-date and harder to implement, such as brownfield redevelopment, more than 10% on-site renewable power generation, etc. Hence, there are a total of 76 items: 27 controlling items, 40 general items and 9 preference items for the evaluation of residential buildings, and there are a total of 83 items: 26 controlling items, 43 general items and 14 for the evaluation of public buildings.

The China green building system grants three levels of ratings: 1-star, 2-star, and 3-star, hence the nickname "Three Star System". The charts below show the different ratings for residential and public buildings.

Problems of green building assessment in China

Though in recent years, the researches of green building assessment tools have made a great progress, there are still many problems need to be solved. And the most important three are followed:

First, for the lack of researches and basis data of assessment like the energy consumption of different materials producing in China, many indicators of the assessment tools are not quantitative. That makes the existed assessment tools not competent for the assessment applications.

Second, with the fast urbanization a lot of big scale projects such as the new town developments emerge in recent years which produce very strong demands for a sustainability assessment tool for big scale residential area development with which the sustainability principles, technologies and methods can be adopted from the initial scheming and planning phrases. Such a tool will enhance the sustainability of the cities to a great extent and provide a better conditional for the efforts on the sustainability on the building scale. But until now, there have been no research on the sustainability assessment tools for big scale residential area development in China.

Third, some assessment tools are not well accepted by the market. Considering the top-down character of economy, politics and society in China, the assessment should find a suitable platform and scale. An unsuitable platform or scale will lead to a lack of practicability, and generally will make the assessment tool not accepted by the market.

3.2 Research of the assessment scale and platform in China

3.2.1 Introduction of Chinese urban planning system and residential development process

Obviously, now there is sufficient impetus to promote the sustainability assessment for big scale residential area development in China. But as mentioned before, the assessment scale, together with the operate platform, is essential for the assessment tool especially in the context of China. A deep research of Chinese urban planning system and residential development process is made here to find out the best assessment scale and platform for the tool.

There has been awareness long ago that the overall design of a residential settlement is a critical determinant of the sustainable development (Jaccard, et al., 1997). While, as the overall design, urban planning has constantly been regarded as the dynamic predictions of city future, which could guide city to develop towards the direction of more sustainable tomorrow. In China, it has become an indispensable means in the field of land use and spatial resource allocation and an effective tool to balance the benefits of social groups and individuals. And it also plays an important role in controlling and regulating the design, construction and management process.

Chinese Urban planning System

Urban planning is a multi-level planning system, and the various levels of planning are under the respective governments' control, with an appropriate level of administrative power in China. This is the so-called "planning in the parallel of government". According to the *Urban and Rural Planning Law of China*, there are four levels of whole planning system: the Urban System Plan, Master Plan, Regulatory Plan and Site Plan. Generally the scope of Urban System Planning is much bigger than a city or metropolis, it is not made and controlled at city level. Generally, the Chinese municipal governments are responsible for the Master plan, Regulatory plan and Site plan of administrative division respectively.

Master plan is a framework for the design, deployment and implementation of measures within a certain period of time (generally in 20 years) on the land use, spatial layout, city infrastructures, etc. It is made by the local governments based on the Urban National Economic and Social Development Plan together with the natural environment, resources condition, history and status quo. As a strategic planning, it is necessary to determine the goals and scales of urban development in next few years, which connect with economic development goals closely. It is the first step of establishing a planning system for a city, creating the development structure of a city from macro viewpoint and providing guides for the construction of the city.

Corresponding to the administration levels, the master plans are approved by different level of governments according to the city's importance and scales. The master plans of large or important cities have to be approved by the state council, while those medium and small cities' master plan must be approved by the provincial governments.

Based on the master planning, regulatory detailed plan or regulatory plan is to determine the construction land use with a set of indicators such as the intensity of use, position of roads and utilities engineering, regulation for space and environment and other special requirements. Regulatory plan divide the designated area into small blocks with a set of indicators to control the population density and building volume and quantitative and qualitative requirements on other specialty area such as municipal services, transport facilities, etc.

Regulatory plan is considered as an interim step of the urban planning system, which could transform the policy and strategic requirements in the master plan into the executable indicators or requirements to control the land exploitation practice. Once the regulatory plans are approved by the municipal governments, it will become obliged ordinances which play an important role in the process of land market and development. It will be discussed in latter.

Based the Master plan and Regulatory plan, Site plan or construction detailed plan is the scheme to guide the building or construction design in the blocks. Site plan should consist of seven aspects as follow:

1. Analysis for site condition and technical and economic feasibility
2. Layout of buildings, roads, green space, etc. and landscape design
3. Planning and design for transport system within the planning blocks
4. Planning for green space, water system and other environment elements within the planning blocks
5. Engineering pipeline comprehensive planning within the planning blocks
6. Vertical planning within the planning blocks
7. Estimation for the whole investment and benefit of the projects

Site plan is the bottom level of whole urban planning system in China. Different from the upper two levels of planning, it is often made by the planning and design institutes commissioned by the land or building development companies rather than the governments. However, it needs the permission to implement from municipal government.

As an administration dominant system, today the urban planning system still operates as a top-down structure in China. Generally speaking, the urban planning system has a much strong power in controlling and regulating the urban development and building construction. Besides, similar with the concept of planning in the parallel of government, the scale of development corresponds the operating platform. For the research scope of this research, it means that once the assessment platform is determined the scale is also determined and vice versa.

Chinese urban residential development

As a main part of urban development, it is doubtless that residential developments follow the procedure mentioned above. However, there are still some features which should be pointed out. Different from other development, in China the residential development has a special hierarchical structure. In China, according to the National code for urban residential district planning &

design, the residential development can be rated into three levels: residential district, residential quarter and neighborhood. They each have their respective scale, population and requirements of municipal services. The national code defines them as follow:

Residential district refers to those residential settlements which are surrounded by arterial roads or natural boundaries and generally have a population of 30,000−50,000 or more.

Residential quarter refers to those residential settlements which are surrounded by urban streets or natural boundaries and generally have a population of 10,000−15,000.

Neighborhood refers to those residential settlements which are surrounded by residential quarter roads and generally have a population of 1,000−3,000.

The residential district and residential quarter developments usually correspond to the second and third step of procedure of Chinese urban land market. The raw land is comprehensively developed into full-serviced urban residential district by municipal governmental companies. At this step, the urban residential district is divided into blocks by the urban streets or natural boundaries, building floor area allowed and some other important requirements of the blocks were written into the regular plan. Then, land use right of the blocks are sold to those building development companies to develop residential quarters with buildings according to the regular plan. And at last, the buildings are sold to the individuals with land use right of part area of the block.

3.2.2 Study of the suitable scale and platform of sustainable residential district assessment

After the research and analysis mentioned before, a hierarchical structure which combines the urban planning system, land market and urban residential development can be established (Fig 3.1).

Just as we mentioned before, in China there have been some index systems working as quantitative indicators and they measure or set the target for sustainable development on the city scale in the eco-city practices. On the other hand, on the building scale some green building assessment tools such as the GOBAS, Three Star System or some globe tools like LEED, BREEAM have been implemented in the residential quarter developments. But until now, the assessment tool for the area scale (residential district scale), which can connect the city and building scales, has been still absent.

In this research, the residential district and regulatory plan has been chosen as the implement

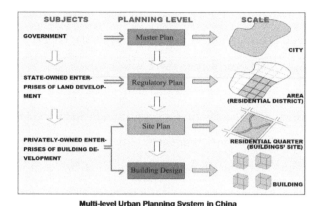

Fig 3.1 The multi-level Urban Planning System and Development in China

scale and platform for several reasons as follow: First of all, in Chinese residential development system residential district is an integrated unit both in the function and structure, which means many important sustainable actions or measures over block scale, such as enhancing accessibility of public transport, creating ecosystem networks or improving sustainable community residents structure can be employed on this level. Second, because the regulatory planning has powerful practicality and operability, an assessment tool based on it can integrate and control the scheme, design, construction and management process efficiently, which can ensure that the sustainable target can be accomplished successfully. Third, such an assessment tool on residential district can provide a better and more equitable condition for green building construction to different building development companies in a same area and reduce the low-price low-performance race between them because of its integrated considerations, assessments and requirements. Forth, the residential district scale and regulatory planning platform can make the assessment tool cooperate better with the assessment tool for building scale—the Evaluation Standard for Green Building without any technical conflicts.

Then, after choosing the implement scale and platform, a research on the assessment indicators will be made as the further step.

Chapter Four: Establishment Process of Sustainable Residential District Assessment Tool

第4章 我国大型城镇住区可持续性评估工具的建立过程

本章致力于探究可持续大型城镇住区评估的核心理念，并剖析其在我国可持续发展战略中的重要地位。通过系统性的分析，力求构建一个既切实可行又具有高效能的可持续居住区评估项目框架，以适应中国独特的城市化进程和社会经济背景。以此框架为基础，研究进一步开发一套较为全面的可持续居住区（SRD）评级体系，以客观量化住区在可持续发展多维指标上的实际表现，进而推动我国城镇住区开发朝着更为绿色、宜居和可持续的方向迈进。

研究通过融合实地调查、问卷调查及访谈等定量与定性相结合的手段，构建了一套综合全面的可持续居住区评估工具，其包含以下 4 个核心步骤：首先，制定评估项目的初步构想与基础框架；其次，确立科学的抽样策略，并执行详细的问卷调查程序；再次，依据所收集的数据进行深入分析，进而开发出有效的评估模型；最后，整合前述研究成果，建立系统完备的评估表和评分机制。通过上述步骤的有序实施，提升住区可持续评估的科学性与实用性。

研究通过系统比较和总结发达国家现有的用于大规模住区的可持续性评估工具，确立了问卷所采用的评估项目的初步框架。为确保评估项目成为实用且有效的工具，应特别注重其准确性、可测量性、可用性和效率。经过深入分析和综合考量，初步构建了一个包含 11 个评估类别、35 个具体评估指标以及 57 个详细评估参数的全面框架。这一框架为后续的可持续性评估提供了坚实的方法论基础和操作指南。之后，研究结合我国的具体国情和实际情况，构建了一套较为合理的权重体系，以明确每个评估类别、指标及参数在可持续居住区评估中的重要程度，并采用加权系数的方法，对各级考核项目进行加权处理。研究进一步开发了一套评估系统，涵盖了经过筛选与权衡的评估类别、指标、参数以及相应的加权系数，确保了评估的全面性与准确性。在此研究框架内，该评估系统被构建成为一种基于标准的评估工具，用于对一系列具有一定规模的住区项目进行客观评分，并通过可视化的程序展示包含排名和类别得分的雷达图等总体评估成果，以及每个评估类别的详细结果和得分表，为住区可持续性的综合评估提供一个全面、高效且操作便捷的计算工具。

研究成果可以转化为针对可持续居住区的评估手册，其由两大部分组成：其一为工具概述部分，旨在为使用者提供该评估工具的基础性认识与理解，涵盖其目的、功能及应用范围等方面；其二则为评估参数详解部分，这一部分的内容更为深入与具体，它不仅为用户提供了关于各个评估参数的额外信息，还详细阐述了如何针对特定参数进行有效的评估操作。通过这两部分的有机结合，本评分系统既能够满足使用者对评估工具的整体把握需求，又能够提供较为细致的参数评估指导，从而确保评估工作的科学性与准确性。

可持续居住区评估系统致力于成为我国在大型城镇住区可持续发展领域的一个实用且高效的评估工具。该系统旨在辅助房地产开发商、城市规划师、设计师以及政府管理机构等相关方优化现有措施，进而证实我国住区在大规模开发过程中的可持续性。

4.1 Introduction of research design

This research attempts to make a better understanding of the concept of sustainable urban residential district assessment and its role for sustainable development in China. Moreover, the research tries to establish a framework of feasible and effective sustainable residential district assessment items for China. Then, based on the framework this research will develop a Sustainable Residential District (SRD) Rating System (a computer program) to measure the performance on the sustainability issues, and thus to promote the sustainable development in the urbanization of China.

The establishment process of the sustainable residential district assessment tool is based on three assumptions.

（1）Developing the sustainable residential district assessment tool should be based on the analysis and comparison of the existing famous assessment systems for the similar scope in the developed countries.

（2）The assessment tool should be suitable for Chinese context.

（3）The assessment tool should focus on the residential district only. Other types of urban areas such as industrial areas are not included in this research.

（4）The developed system should consider the implementation in the procedure of Chinese residential development, relational contexts such as the national standard, codes, status quo, etc. should be considered.

Then, a multi-dimensional design strategy that involves many quantitative and qualitative approaches such as field survey, questionnaire and interviews is adopted in this research to establish the sustainable residential district assessment tool. A four-step methodology is designed as follow.

1. Preparation of the initial framework of assessment items

Comparison and analysis of existing famous sustainability assessment tools for residential district level is made to determine the initial framework of assessment items which will be used in the questionnaire. In order to establish a comprehensive and feasible framework, the review will compare the assessment items of existing different systems according to a same categories classification which will also be used in the suggested assessment tool developed by the research.

Then based on the context of China an initial framework of assessment items would be established. All the assessment items which are considered to be positive for the sustainable assessment are included in the initial framework to ensure its comprehensiveness.

2. Sampling and questionnaire procedure

The sample will include a group of stakeholders from different fields: urban planning and architectural design, engineering, construction management offices, estate developers, students and others. All the participants are classified into two main groups. The first group consists of experts and professionals. The second group is composed of laymen who are undergraduate students concerned on the sustainability issue. They are selected according to their role and influence on sustainable development practices. The participants were composed of (50%) experts and (50%) laymen (non-professionals) with a total number of 60.

Then the two groups of respondents are asked to answer the questionnaire which is prepared in the first step, which is composed of a series of open and closed questions. All the respondents will be asked to assess all the assessment parameter in three aspects (Understandability, Utility and Practicability) and then delete assessment items which are not important for the tool or infeasible in the assessment practice and make suggestions to add new assessment items which they think are necessary for the assessment tool.

3. Data analysis and evaluation model developing procedure

Because the sustainable residential district assessment tool is a multi-dimensional method involving different environmental, social, and economical issues, the process of questionnaire should adopt different integrated methodologies such as Experts panel, Endpoint method, Economy method, AHP method, and others. After considering the advantages and disadvantages of each method to build a new compatible tool and the purposes of this study, the AHP method is finally adopted by this research.

After the questionnaire results of the second step are collected and analyzed to define the final framework of assessment items, some respondents of the expert group will answer the questionnaire made by the AHP software Expert Choice. By using this technique, the researcher can identify the relative importance of the categories, indicators, and parameters of the assessment system with respect to the overall target: a sustainable residential district and their weightings. Now it is well accepted that this method of data collection and analysis can help the researchers and decision makers find an alternative which satisfies their goal and understand the problems better.

The Analytic Hierarchy Process (AHP) is a structured technique for organizing and analyzing complex decisions. Based on mathematics and psychology, it was developed by Thomas L. Saaty in the 1970s and has been extensively studied and refined since then (Wikipedia). This method allows consideration of both qualitative and quantitative aspects and can reduce complex decisions to a series of pair comparisons with weighting criteria and sub-criteria, analyzing the data and finding in the decision process. In addition, AHP helps in reducing bias in decision-making process, such as lack of focus, planning, etc., which ultimately leads to distractions and makes the teams cannot make the right choice (H. Ali, et al. 2008).

The basic concepts of AHP methodology are decomposition, pair-wise comparison, and synthesis of priorities. For the first step, the user of AHP should decompose the problem into a hierarchical structure of less comprehensive sub-problems which will be analyzed independently according to its main components. Then for the second step, the decision makers will assess every element by comparing every pair of elements' impact on the target, a criterion or element above their level in the hierarchy at a time to derive their weights or priorities. And then, base on the evaluation a numerical weight or priority of each element in the hierarchy is derived. Finally, the numerical priorities of decision alternatives are synthesized into an overall rating to make the best decision.

Finally, the pair-wise comparison results of the final parameters, indicators and categories will be analyzed by the AHP software Expert Choice to set up the weighting coefficient system.

4. Assessment sheets and rating system set-up

Based on the previous findings that included the final assessment categories, evaluation indicators, and evaluation parameters and their weighting coefficient, an assessment program based on the Excel sheets was developed. This system is classified as an application-oriented tool that assigns point values to a selected number of assessment items on a certain scale ranging among five levels, and the middle level is the average performance or the national standards.

The computer based program was developed to calculate the overall level of sustainability performance of the assessment residential district project. The excel spread sheets consist of the main sheet that includes basic information of the assessment project, the results sheet that includes summary of the designated area, overall score level and assessment of indicators' results of each category, the score sheet that includes all the score judgments of the assessment parameters, together with the weight of assessment items respectively. Moreover, the program has 11 assessment sheets of the following categories: Land use & Site, Energy, Water, Materials, Solid waste & management, Ecology, Transport, Outdoor environment & climate, Community Economy & Sociability, Infrastructure and Building. Each sheet of category includes its indicators and parameters with their weights derived from the AHP method by using the Expert Choice software. The score of each parameter comes from multiplying the points of each parameter (0~4 points are used to indicate the level of sustainability for each parameter) by their specific weight with respect to their indicator. The total sum of the parameters will appear on their indicator level and then multiplied by the indicator's weight with respect to their category. Finally the overall score comes from the score of each category multiplied by their relative weights with respect to the top target: A sustainable residential district. These values will be shown on the results and score sheet that indicates the contribution of each category and the summation of all weights. These results will be presented graphically showing the overall level of the sustainability.

The rating criterion is based on the summation of existing assessment system. The score for each assessment item is multiplied by the weighting coefficient, and aggregated into summation. The final rating depends on the overall score percentage. According to the analyses of the developed rating systems such as LEED and BREEAM, a five performance levels rating system is accepted, which includes Excellent (100%–85%), Good (85%–70%), Satisfactory (70%–50%), Pass (50%–30%), Unclassified (<30%).

4.2 Preparation of the initial framework of assessment items and questionnaire

4.2.1 Introduction

The assessment items are crucial for the sustainable residential district assessment tool. In the evaluation process, the assessment items are the instruments to measure how sustainable a residential district project is, because they can simplify and quantify the information of different aspects. So as the first step, an initial framework which includes all the useful assessment items should be established. The selection of assessment items should follow these requirements.

(1) The assessment items should be comprehensive in evaluating the sustainability performance of residential district from different aspects of environment, society, and economy.

(2) The assessment items should be representative of sustainable development's main concepts and principles.

(3) The assessment items should acknowledge the local context of China and consider the feasibility in assessment practice.

(4) The assessment items should consider the suitable assessment stages which could cooperate with the process of Chinese residential district development.

In addition, in order to be a useful and feasible instrument, the assessment items should have some necessary features as follow.

(1) Accuracy: The assessment items can represent the elements needed to be measured, without any ambiguity.

(2) Measurability: The assessment items should make the quality possible to be measured.

(3) Availability: The assessment items should consider the possibility and feasibility of getting the reliable data or information in the local context of China.

(4) Efficiency: The assessment items should be clear, simple and easy for comprehension and communication.

It should be pointed out that the assessment items of existing tools often have a hierarchical structure of 2~3 levels. BREEAM Communities and CASBEE for Urban Development have a three-level structure, while LEED for Neighborhood Development has a two-level one. No matter how many levels the structure has the basic level is the core because it directly represents the particular measures or actions. In order to make the assessment items comparison and analysis between different tools, it is necessary to unify the names of different assessment item levels. For the convenience's sake, in this research the items of the basic level would be named as parameters while those on the topic level would be named as categories. The items on the mid-level would be named as indicators. In this research such a 3 level structure would be adopted both in the initial and final framework of assessment items of sustainable residential district assessment tool for China.

4.2.2 The assessment items comparison and analysis of existing sustainability assessment tools for area scale in developed countries

As mentioned in Chapter 1, until now there have been three famous existing assessment tools for area scale in developed countries, which are BREEAM Communities, LEED for Neighborhood Development and CASBEE for Urban Development. All these three assessment tools have a very comprehensive framework of assessment items with technical details, which makes them quite referential for this research and worthy to make a deep comparison and analysis before establishing the initial framework of assessment items of the sustainable residential district assessment tool for China.

The assessment items of BREEAM Communities

BREEAM Communities has a three level structure of assessment items. The top level is sections (category level) which are:

（1）Climate & energy

（2）Place shaping

（3）Community

（4）Transport & movement

（5）Ecology and biodiversity

（6）Resources

（7）Business and economy

（8）Buildings

Then, each of them has a series of sub-categories (indicator level) which consist of a number of issues (parameter level) to define the sustainable target and the evaluation criteria and sub-criteria for the assessment projects.

The introduction of the three-level framework of assessment items of BREEAM Communities is shown in Table 4.1.

Table 4.1 The framework of assessment items of BREEAM Communities (BREEAM, 2009)

Sections (Categories)	Sub-categories (Indicators)	Issues (Parameters)
Climate & energy	Water management	CE1 – Flood risk assessment
		CE2 – Surface water runoff
		CE3 – Rainwater SUDS
	Design principles	CE4 – Heat island

Sections (Categories)	Sub-categories (Indicators)	Issues (Parameters)
	Energy management	CE5 – Energy efficiency
		CE6 – Onsite renewable(s)
		CE7 – Future renewable(s)
	Infrastructure	CE8 – Services
	Water resources management	CE9 – Water consumption
Place shaping	Effective use of land	PS1 – Sequential approach
		PS2 – Land reuse
		PS3 – Building reuse
	Design process	PS4 – Landscaping
		PS5 – Design and access
	Open space	PS6 – Green areas
	Inclusive communities	PS7 – Local demographics
		PS8 – Affordable housing
	Form of development	PS9 – Secure by design
		PS10 – Active frontages
		PS11 – Defensible spaces
Community	Inclusive communities	COM1 – Inclusive design
		COM2 – Consultation
		COM3 – Development user guide
		COM4 – Management and operation
Transport & movement	Public transport	TRA 1 – Location / Capacity
		TRA 2 – Availability / Frequency
		TRA 3 – Facilities
	General policy	TRA 4 – Local amenities
	Cycling	TRA 5 – Network
		TRA 6 – Facilities
	Traffic	TRA 7 – Car clubs
		TRA 8 – Flexible parking
		TRA 9 – Local parking
		TRA 10 – Home zones
		TRA 11 – Transport assessment

Sections (Categories)	Sub-categories (Indicators)	Issues (Parameters)
Ecology and biodiversity	Ecology and biodiversity	ECO1 – Ecological survey
		ECO2 – Biodiversity action plan
		ECO3 – Native flora
Resources	Materials	RES 1 – Low impact
		RES 2 – Locally sourced materials
		RES 3 – Road construction
	Waste operation	RES 4 – Composting
	Water resources	RES 5 – Water master planning strategy
		RES 6 – Groundwater
Business and economy		Bus 1 – Business priority sectors
		Bus 2 – Labor and skills
		BUS 3 – Employment
		BUS 4 – New business
		BUS 5 – Investment
Buildings	Code for sustainable homes / EcoHomes	BLD 1 – Domestic
	BREEAM (or equivalent)	BLD 2 – Non domestic

LEED for Neighborhood Development

The LEED for Neighborhood Development is a rating system for neighborhood planning and development, which combines the principles of smart growth, New Urbanism, and green building into the national sustainability assessment system for the first time. Different from the assessment items of BREEAM Communities and CASBEE for Urban Development, LEED for Neighborhood Development has prerequisite which must be satisfied and credits in the rating system addressing five topics:

（1）Smart location and linkage (SLL)

（2）Neighborhood pattern and design (NPD)

（3）Green infrastructure and buildings (GIB)

（4）Innovation and design process (IDP)

（5）Regional priority credit (RPC)

The introduction of the two-level framework of assessment items of LEED for Neighborhood Development is shown in Table 4.2.

Table 4.2 The framework of assessment items of LEED for Neighborhood (LEED-ND, 2009)

Topics	Prerequisite and credits
Smart location and linkage (SLL)	prerequisite 1 smart location
	prerequisite 2 Imperiled species and ecological communities
	prerequisite 3 Wetland and water body conservation
	prerequisite 4 Agricultural land conservation
	prerequisite 5 Floodplain avoidance
	Credit 1 Preferred locations
	Credit 2 Brownfield redevelopment
	Credit 3 Locations with reduced automobile dependence
	Credit 4 Bicycle network and storage
	Credit 5 Housing and jobs proximity
	Credit 6 Steep slope protection
	Credit 7 Site design for habitat or wetland and water body conservation
	Credit 8 Restoration of habitat or wetlands and water bodies
	Credit 9 Long-term conservation management of habitat or wetlands and water bodies
Neighborhood pattern and design (NPD)	prerequisite 1 Walkable streets
	prerequisite 2 Compact development
	prerequisite 3 Connected and open community
	Credit 1 Walkable streets
	Credit 2 Compact development
	Credit 3 Mixed-use neighborhood centers
	Credit 4 Mixed-income diverse communities
	Credit 5 Reduced parking footprint
	Credit 6 Street network
	Credit 7 Transit facilities
	Credit 8 Transportation demand management
	Credit 9 Access to civic and public spaces
	Credit 10 Access to recreation facilities
	Credit 11 Visitability and universal design
	Credit 12 Community outreach and involvement
	Credit 13 Local food production
	Credit 14 Tree-lined and shaded streets
	Credit 15 Neighborhood schools

Topics	Prerequisite and credits
Green infrastructure and buildings (GIB)	prerequisite 1　Certified green building
	prerequisite 2　Minimum building energy efficiency
	prerequisite 3　Minimum building water efficiency
	prerequisite 4　Construction activity pollution prevention
	Credit 1　Certified green buildings
	Credit 2　Building energy efficiency
	Credit 3　Building water efficiency
	Credit 4　Water-efficient landscaping
	Credit 5　Existing building reuse
	Credit 6　Historic resource preservation and adaptive use
	Credit 7　Minimized site disturbance in design and construction
	Credit 8　Stormwater management
	Credit 9　Heat island reduction
	Credit 10　Solar orientation
	Credit 11　On-site renewable energy sources
	Credit 12　District heating and cooling
	Credit 13　Infrastructure energy efficiency
	Credit 14　Wastewater management
	Credit 15　Recycled content in infrastructure
	Credit 16　Solid waste management infrastructure
	Credit 17　Light pollution reduction
Innovation and design process (IDP)	Credit 1　Innovation and exemplary performance
	Credit 2　LEED® accredited professional
Regional priority credit (RPC)	Credit 1　Regional priority

CASBEE for Urban Development

CASBEE uses a benchmark measurement system which is different from other assessment systems. According to the concept of CASBEE, the closed ecosystem is essential for determining environment assessments. Such concept is also carried on and applied by CASBEE for Urban Development. In order to define the term Q (Quality: environmental quality) and L (Load: building environmental load) used in the *BEE* definition, the hypothetical boundary is set around

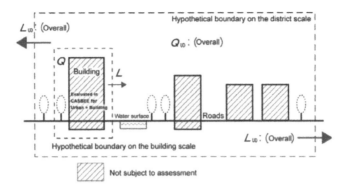

Fig 4.1 Definition of Q and L through Hypothetical Boundary (CASBEE, 2007)

the assessment area development project (Fig 4.1). Then, the assessment tool addresses both the environmental quality within the hypothetical boundary (Q_{UD}: Mainly applying to the Q3 field on the building scale) and environmental load (L_{UD}: Mainly applying to the L3 field on the building scale) beyond that boundary. Because the buildings have been assessed by the previous CASBEE (building scale) edition, they are out of the assessment boundary of CASBEE for Urban Development. Another characteristic of CASBEE is that it assigns separate scores for Q and L and ultimately gives an assessment of BEE as an indicator based on the results for Q and L. L is firstly evaluated as LR (environmental load reduction of the building). That approach is employed because "higher marks for improving load reduction quality" is easier to understand than "higher marks for load reduction" as an assessment system, just as "improvements in quality and performance earn higher marks." Such assessment way is also adopted by CASBEE for Urban Development. Q_{UD} (environmental quality in urban development) and L_{UD} (outdoor environmental loads in urban development), each comprises three main categories, are assessed and scored independently. And then, all the six categories are also compounded using the formula below to generate BEE_{UD}, an indicator for Building Environmental Efficiency in urban development.

$$BEE_{UD} = \frac{Q_{UD}}{L_{UD}}$$

Where, L_{UD} is first evaluated as LR_{UD} (Load Reduction in urban development).

The assessment items for Q_{UD} and LR_{UD} are then divided into three categories each, and each category comprise 4~6 sub-categories (indicators), and then each sub-category is further divided into minor-categories (parameters). Each minor-category is scored on five levels, according to predetermined criteria, and weighting coefficients are applied between assessment fields to calculate the results, which is also the same approach as CASBEE (building scale).

All the assessment items of CASBEE for Urban Development are shown in Table 4.3 & Table 4.4.

Table 4.3 Assessment items included in Q_{UD} of CASBEE for Urban Development (CASBEE, 2007)

Categories	Indicators	Parameters
Q_{UD}1 Natural Environment (microclimates and ecosystems)	1.1 Consideration and conservation of microclimates in pedestrian space in summer	1.1.1 Mitigation of heat island effect with the passage of air
		1.1.2 Mitigation of heat island effect with shading
		1.1.3 Mitigation of heat island effect with green space and open water etc.
		1.1.4 Consideration for the positioning of heat exhaust
	1.2 Consideration and conservation of terrain	1.2.1 Building layout and shape design that consider existing topographic character
		1.2.2 Conservation of topsoil
		1.2.3 Consideration of soil contamination
	1.3 Consideration and conservation of water environment	1.3.1 Conservation of water bodies
		1.3.2 Conservation of aquifers
		1.3.3 Consideration of water quality
	1.4 Conservation and creation of habitat	1.4.1 Grasping the potential of the natural environment
		1.4.2 Conservation or regeneration of natural resources
		1.4.3 Creating ecosystem networks
		1.4.4 Providing a suitable habitat for flora and fauna
	1.5 Other consideration for the environment inside the designated area	1.5.1 Ensuring good air quality, acoustic and vibration environments
		1.5.2 Improving the wind environment
		1.5.3 Securing sunlight
Q_{UD}2 Service functions for the designated area	2.1 Performance of supply and treatment systems (mains water, sewerage and energy)	2.1.1 Reliability of supply and treatment systems
		2.1.2 Flexibility to meet changing demand and technical innovation in supply and treatment systems

Categories	Indicators	Parameters
	2.2 Performance of information systems	2.2.1 Reliability of information systems
		2.2.2 Flexibility to meet changing demand and technical innovation in information systems
		2.2.3 Usability
	2.3 Performance of transportation systems	2.3.1 Sufficient capacity of transportation systems
		2.3.2 Securing safety in pedestrian areas etc.
	2.4 Disaster and crime prevention performance	2.4.1 Understanding the risk from natural hazards
		2.4.2 Securing open space as wide area shelter
		2.4.3 Providing proper evacuation routes
		2.4.4 Crime prevention performance (surveillance and territoriality)
	2.5 Convenience of daily life	2.5.1 Distance to daily-use stores and facilities
		2.5.2 Distance to medical and welfare facilities
		2.5.3 Distance to educational and cultural facilities
	2.6 Consideration for universal design	
$Q_{UD}3$ Contribution to the local community (history, culture, scenery and revitalization)	3.1 Use of local resources	3.1.1 Use of local industries, personnel and skills
		3.1.2 Conservation and use of historical, cultural and natural assets
	3.2 Contribution to the formation of social infrastructure	
	3.3 Consideration for nurturing a good community	3.3.1 Formation of local centers and fostering of vitality and communication
		3.3.2 Creation of various opportunities for public involvement
	3.4. Consideration for urban context and scenery	3.4.1 Formation of urban context and scenery
		3.4.2 Harmony with surroundings

Table 4.4 Assessment items included in LRUD of CASBEE for Urban Development (CASBEE, 2007)

Categories	Indicators	Parameters
$LR_{UD}1$ Environmental impact on microclimates, façade and landscape	1.1 Reduction of thermal impact on the environment outside the designated area in summer	1.1.1 Planning of building group layout and forms to avoid blocking wind
		1.1.2 Consideration for paving materials
		1.1.3 Consideration for building cladding materials
		1.1.4 Consideration for reduction of waste heat
	1.2 Mitigation of impact on geological features outside the designated area	1.2.1 Prevention of soil contamination
		1.2.2 Reduction of ground subsidence
	1.3 Prevention of air pollution affecting outside the designated area	1.3.1 Source control measures
		1.3.2 Measures concerning means of transport
		1.3.3 Atmospheric purification measures
	1.4 Prevention of noise, vibration and odor affecting outside the designated area	1.4.1 Reduction of the impact of noise
		1.4.2 Reduction of the impact of vibration
		1.4.3 Reduction of the impact of odor
	1.5 Mitigation of wind hazard and sunlight obstruction affecting outside the designated area	1.5.1 Mitigation of wind hazard
		1.5.2 Mitigation of sunlight obstruction
	1.6 Mitigation of light pollution affecting outside the designated area	1.6.1 Mitigation of light pollution from lighting and advertising displays etc.
		1.6.2 Mitigation of sunlight reflection from building facade and landscape materials
$LR_{UD}2$ Social infrastructure	2.1 Reduction of mains water supply (load)	2.1.1 Encouragement for the use of stored rainwater
		2.1.2 Water recirculation and use through a miscellaneous water system
	2.2 Reduction of rainwater discharge load	2.2.1 Mitigation of surface water runoff using permeable paving and percolation trenches
		2.2.2 Mitigation of rainwater outflow using retaining pond and flood control basins
	2.3 Reduction of the treatment load from sewage and gray water	2.3.1 Load reduction using high-level treatment of sewage and gray water
		2.3.2 Load leveling using water discharge balancing tanks etc.

Categories	Indicators	Parameters
	2.4 Reduction of waste treatment load	2.4.1 Reduction of collection load using centralized-storage facilities
		2.4.2 Installation of facilities to reduce the volume and weight of waste and employ composting
		2.4.3 Classification, treatment and disposal of waste
	2.5 Consideration for traffic load	2.5.1 Reduction of the total traffic volume through modal shift
		2.5.2 Efficient traffic assignment on local road network
	2.6 Effective energy use for the entire designated area	2.6.1 Area network of unused and renewable energy
		2.6.2 Load leveling of electrical power and heat through area network
		2.6.3 Area network of high-efficient energy system
$LR_{UD}3$ Management of the local environment	3.1 Consideration of global warming	3.1.1 Construction and materials, etc.
		3.1.2 Energy
		3.1.3 Transportation
	3.2 Environmentally responsible construction management	3.2.1 Acquisition of ISO14001 certification
		3.2.2 Reduction of by-products of construction
		3.2.3 Energy saving activity during construction
		3.2.4 Reduction of construction-related impact affecting outside the designated area
		3.2.5 Selection of materials with consideration for the global environment
		3.2.6 Selection of materials with consideration for impact on health
	3.3 Regional transportation planning	3.3.1 Coordinating with the administrative master plans for transportation system
		3.3.2 Measures for transportation demand management
	3.4 Monitoring and management system	3.4.1 Monitoring and management system to reduce energy usage inside the designated area
		3.4.2 Monitoring and management system to conserve the surrounding environment of the designated area

Comparison and Conclusion

Comparison and conclusion of existing famous sustainability assessment tools for residential district level is necessary for determining the initial framework of assessment items which will be adopted by the questionnaire. In order to make the comparison between different tools a unified comprehensive categories classification which covers all the assessment aspects should be defined firstly. Furthermore, the categories classification will also be used in the assessment tool developed by the research.

After comparison and analysis of the existing assessment tools, a framework of 11 assessment categories are defined as following:

（1）Land & site

（2）Energy

（3）Water

（4）Materials

（5）Solid waste & management

（6）Ecology

（7）Transport

（8）Outdoor environment & climate

（9）Community economy & sociality

（10）Infrastructure

（11）Building

Then, in order to compare the three assessment tools, all the basic level assessment items of three existing assessment tools are rearranged according to the 11 categories as shown in Table 4.5.

Table 4.5 The basic level assessment items comparison of BREEAM Communities, LEED for Neighborhood Development and CASBEE for Urban Development

		BREEAM Communities	LEED for Neighborhood Development	CASBEE for Urban Development
Land & Site		CE1 Flood risk assessment	SLL P5 Floodplain avoidance	Q 1.2.1 Building layout and shape design that consider existing topographic character
		PS1 Sequential approach	SLL C6 Steep slope protection	Q 1.2.2 Conservation of topsoil

	BREEAM Communities	LEED for Neighborhood Development	CASBEE for Urban Development
	PS2 Land reuse	SLL P4 Agricultural land conservation	Q 1.2.3 Consideration of soil contamination
	PS3 Building reuse	SLL C1 Preferred locations	Q 1.2.1 Prevention of soil contamination
		SLL C2 Brownfield redevelopment	Q 1.2.2 Reduction of ground subsidence
		NPD P2 Compact development	
		NPD C2 Compact development	
		GIB C7 Minimized site disturbance in design and construction	
Energy	CE5 Energy efficiency	GIB P2 minimum building energy efficiency	LR 1.5.2 Mitigation of sunlight obstruction
	CE6 Onsite renewable(s)	GIB C2 Building energy efficiency	LR 2.6.1 Area network of unused and renewable energy
	CE7 Future renewable(s)	GIB C10 Solar orientation	LR 2.6.2 Load leveling of electrical power and heat through area network
	CE8 Services	GIB C11 On-site renewable energy sources	LR 2.6.3 Area network of high-efficient energy system
		GIB C12 District heating and cooling	LR 3.1.2 Energy (management of local environment)
		GIB C13 Infrastructure energy efficiency	LR 3.2.3 Energy-saving activity during construction
Water	CE2 Surface water run-off	SLL P3 Wetland and water body conservation	Q 1.3.1 Conservation of water bodies
	CE3 Rain-SUDS	GIB P3 minimum building water efficiency	Q 1.3.2 Conservation of aquifers
	CE9 Water consumption	GIB P4 Construction activity pollution prevention	Q 1.3.3 Consideration of water quality
	RES5 Water master planning strategy	GIB C3 Building water efficiency	LR 2.1.1 Encouragement for the use of stored rainwater
	RES6 Ground water	GIB C4 Water-efficient landscaping	LR 2.1.2 Water recirculation and use through a miscellaneous water system
		GIB C8 Stormwater management	LR 2.2.1 Mitigation of surface water runoff using permeable paving and percolation trenches

	BREEAM Communities	LEED for Neighborhood Development	CASBEE for Urban Development
		GIB C14 Wastewater management	LR 2.2.2 Mitigation of rainwater outflow using retaining pond and flood control basins
			LR 2.3.1 Load reduction using high-level treatment of sewage and graywater
			LR 2.3.2 Load leveling using water discharge balancing tanks etc.
Materials	RES1 Low impact materials	GIB P4 Construction activity pollution prevention	LR 3.1.1 Construction and materials, etc.
	RES2 Locally sourced materials	GIB C5 Existing building reuse	LR 3.2.2 Reduction of by-products of construction
	RES3 Road construction	GIB C15 Recycled content in infrastructure	LR 3.2.5 Selection of materials with consideration for the global environment
	RES4 Composting		LR 3.2.6 Selection of materials with concern for impact on health
Solid waste and management		GIB C16 Solid waste management infrastructure	LR 2.4.1 Reduction of collection load using centralized-storage facilities
			LR 2.4.2 Installation of facilities to reduce the volume and weight of garbage and employ composting
			LR 2.4.3 Classification, treatment and disposal of garbage
Ecology	ECO1 Ecology survey	SLL P2 Imperiled species and ecological communities	Q 1.4.1 Grasping the potential of the natural environment
	ECO2 Biodiversity action plan	SLL C7 Site design for habitat or wetland and water body conservation	Q 1.4.2 Conserving natural resources
	ECO3 Native flora	SLL C8 restoration of habitat or wetlands and water bodies	Q 1.4.3 Creating ecosystem networks
		SLL C9 Long-term conservation management of habitat or wetlands and water bodies	Q 1.4.4 Providing a suitable habitat for flora and fauna
Transport	TRA1 Public transport location/ capacity	SLL P1 Smart location	Q 2.3.1 Sufficient capacity of transportation systems

	BREEAM Communities	LEED for Neighborhood Development	CASBEE for Urban Development
	TRA2 Public transport availability/ frequent	SLL C3 Locations with reduced automobile dependence	Q 2.3.2 Securing safety in pedestrian areas etc.
	TRA3 Public transport facilities	SLL C4 Bicycle network and storage	Q 2.5.1 Distance to daily-use stores and facilities
	TRA4 Local amenities	NPD P1 Walkable Streets	Q 2.5.2 Distance to medical and welfare facilities
	TRA5 Cycling-network	NPD P3 Connected and open community	Q 2.5.3 Distance to educational and cultural facilities
	TRA6 Cycling facilities	NPD C1 Walkable streets	Q 3.1.1 Use of local industries, personnel and skills
	TRA7 Car club	NPD C3 Mixed-use neighborhood centers	LR 2.5.1 Reduction of the total traffic volume through modal shift
	TRA8 Flexible parking	NPD C5 Reduced parking footprint	LR 2.5.2 Efficient traffic assignment on local road network
	TRA9 Local parking	NPD C6 Street network	LR 3.1.3 Transportation
	TRA10 Home zones	NPD C7 Transit facilities	LR 3.3.1 Coordinating with the administrative master plans for transportation system
	TRA11 Transport assessment	NPD C8 Transportation demand management	3.3.2 Measures for transportation demand management
Outdoor environment & climate	CE4 Heat island	NPD C9 Access to civic and public spaces	Q 1.1.1 Mitigation of heat island effect with the passage of air
	PS4 Landscaping	NPD C14 Tree-lined and shaded streets	Q 1.1.2 Mitigation of heat island effect with shading
	PS5 Design and access	GIB C9 Heat Island reduction	Q 1.1.3 Mitigation of heat island effect with green space and open water etc
	PS6 Green areas	GIB C17 Light pollution reduction	Q 1.1.4 Consideration for the positioning of heat exhaust
	PS10 Active Frontages		Q 1.5.1 Ensuring good air quality, acoustic and vibration environments
	PS11 Defensible spaces		Q 1.5.2 Improving the wind environment
			Q 1.5.3 Securing sunlight
			LR 1.1.1 Planning of building group layout and forms to avoid blocking wind

	BREEAM Communities	LEED for Neighborhood Development	CASBEE for Urban Development
			LR 1.1.2 Consideration for paving materials
			LR 1.1.3 Consideration for building cladding materials
			LR 1.1.4 Consideration for reduction of waste heat
			LR 1.3.1 Source control measures
			LR 1.3.2 Measures concerning means of transport
			LR 1.3.3 Atmospheric purification measures
			LR 1.4.1 Reduction of the impact of noise
			LR 1.4.2 Reduction of the impact of vibration
			LR 1.4.3 Reduction of the impact of odor
			LR 1.5.1 Mitigation of wind hazard
			LR 1.6.1 Mitigation of light pollution from lighting and advertising displays etc.
			LR 1.6.2 Mitigation of light reflection from building facade and landscape materials
			LR 3.2.4 Reduction of construction-related impact affecting outside the designated area
Community economy & sociability	COM1 Inclusive design	SLL C5 Housing and jobs proximity	Q 2.4.1 Understanding the risk from natural hazards
	COM2 Consultation	NPD C4 Mixed-income diverse communities	Q 2.4.2 Securing open space as wide area shelter
	COM3 Development user guide	NPD C10 Access to recreation facilities	Q 2.4.3 Providing proper evacuation routes
	COM4 Management and operation	NPD C11 Visitability and universal design	Q 2.4.4 Crime prevention performance (surveillance and territoriality)
	BUS1 Business priority sectors	NPD C12 Community outreach and involvement	Q 2.6 Consideration for universal design

	BREEAM Communities	LEED for Neighborhood Development	CASBEE for Urban Development
	BUS2 Labour & skills	NPD C13 Local food production	Q 3.1.2 Conservation and use of historical, cultural and natural assets
	BUS3 Employment	NPD C15 Neighborhood Schools	Q 3.3.1 Formation of local centers and fostering of vitality and communication
	BUS4 New business	GIB C6 Historic resource preservation and adaptive use	Q 3.3.2 Creation of various opportunities for public involvement
	BUS5 Investment		Q 3.4.1 Formation of urban context and scenery
	PS7 Local demographics		Q 3.4.2 Harmony with surroundings
	PS8 Affordable housing		LR 3.2.1 Acquisition of ISO14001 certification
	PS9 Secure by design		3.4.1 Monitoring and management system to reduce energy usage inside the designated area
			3.4.2 Monitoring and management system to conserve the surrounding environment of the designated area
Infrastructure			Q 2.1.1 Reliability of supply and treatment systems
			Q 2.1.2 Flexibility to meet changing demand and technical innovation in supply and treatment systems
			Q 2.2.1 Reliability of information systems
			Q 2.2.2 Flexibility to meet changing demand and technical innovation in information systems
			Q 2.2.3 Usability
			Q 3.2 Contribution to the formation of social infrastructure
Buildings	BD1 Buildings (Domestic)	GIB P1 Certified green building	
	BD2 Buildings (non-domestic)	GIB C1 Certified green buildings	

4.2.3 The initial framework of assessment items

Based on comparison and analysis of the three existing sustainability assessment tools for the residential district scale in developed countries and considering the context of China, the research defined a three-level hierarchy of assessment items as the initial framework which comprises categories, indicators and parameters. As the top level of the framework, all of the eleven categories will be identified by a group of indicators which varies from one category to another according to the category itself and the local context of China. Also, each indicator is defined through a number of parameters. The initial framework of assessment items are shown in Table 4.6.

Table 4.6 The initial framework of assessment items of the Sustainable Residential District Assessment Tool

N°	Initial categories		Initial indicators		Initial parameters	Qualitative/ Quantitative
1	Land & site	In1	Site selection	ls1	Retrofit area of land used	QN
2				ls2	Impact of the development on the landscape	QL
3		In2	Compact development	ls3	The density of residential development	QN
4				ls4	Site coverage	QN
5		In3	Soil conservation	ls5	Environmental recovery of the excavated soil	QN
6				ls6	Soil permeability	QN
7	Energy	In4	Usage of sunlight	en1	Mitigation of sunlight obstruction requirement	QL
8		In5	Renewable energy	en2	Energy efficiency in the public space	QN
9				en3	Effective results of renewable energy	QN
10		In6	Energy-saving during construction	en4	Evaluate energy-saving activities at the construction	QL
11		In7	Decrease the primary energy demand	en5	Thermal efficiency level	QN
12				en6	Electrical efficiency level	QN

N°	Initial categories	Initial indicators		Initial parameters		Qualitative/ Quantitative
13	Water	In8	Conservation of water bodies	w1	Conservation area of water bodies	QN
14				w2	Consideration of water quality	QL
15		In9	Conservation of aquifers	w3	Groundwater conservation	QL
16		In10	Water use	w4	Usage of unconventional water resources	QN
17		In11	Water-saving infrastructure in public space	w5	Water consumption saving of the infrastructures	QN
18	Materials	In12	Locally sourced materials in public space	m1	Usage of locally sourced materials in public space	QN
19		In13	Recycled and waste resourced materials in public space	m2	Usage of recyclable materials in public space	QN
20				m3	Usage of materials made from waste in public space	QN
21	Solid waste & management	In14	Waste equipment plan	sw1	Waste collection equipments plan in public space	QN
22				sw2	Requirements of waste collection equipments plan in blocks	QN
23		In15	Waste management	sw3	Classification, treatment and disposal of waste	QL
24		In16	Reduction of by-products of construction	sw4	Reducing the generation of construction by-products	QL
25				sw5	Sorting and recycling of construction by-products	QL
26	Ecology	In17	Green land	ec1	Percentage of green land	QN
27		In18	Conserving natural resources	ec2	Conservation of natural resources	QL
28				ec3	Consideration of natural flora in the surroundings	QL
29		In19	Creating ecosystem networks	ec4	Eco mosaic fragmentation	QL

N°	Initial categories		Initial indicators		Initial parameters	Qualitative/ Quantitative
30	Transport	In20	Capacity of transportation systems	t1	Sufficient capacity of transportation systems	QL
31		In21	Accessibility of public transport	t2	Accessibility of public transport	QN
32		In22	Locations with reduced automobile dependence	t3	Convenience of commercial concentrations	QN
33				t4	Convenience of financial institution (including ATM)	QN
34				t5	Convenience of administrative offices	QN
35				t6	Convenience of medical facility	QN
36		In23	Pedestrian traffic	t7	Securing safety in pedestrian area	QL
37		In24	Green travel measures	t8	Existence of green travel measures	QL
38	Outdoor environment & climate	In25	Mitigation of heat island	o1	Horizontal shaded area ratio	QN
39				o2	Open space ratio	QN
40		In26	Improving acoustic environment	o3	Measures to reduce the impact of noise	QL
41		In27	Improving wind environment	o4	Measures to improving wind environment	QL
42		In28	Improving outdoor public space environment	o5	Consideration for the positioning of heat exhaust	QN
43				o6	Vertical planting	QN
44	Community economy & sociality	In29	Sustainable community residents structure	c1	Proportion of affordable residence for low income families and residence with low rent	QN
45		In30	Sustainable community security	c2	Universal design and logo	QL
46				c3	Measures of crime prevention	QL
47				c4	Securing open space as wide area shelter and proper evacuation routes	QL

N°	Initial categories		Initial indicators	Initial parameters		Qualitative/ Quantitative
48			In31 Sustainable community culture	c5	Formation of local centers & fostering of vitality & communication	QL
49				c6	Community involvement	QL
50				c7	Conservation and use of historical and cultural assets	QL
51			In32 Sustainable community management	c8	Acquisition of ISO 14001 certification	QL
52				c9	Formation of urban context and scenery	QL
53	Infrastructure		In33 Performance of Information systems	i1	Reliability of information systems	QL
54				i2	Flexibility to meet changing demand and technical innovation in information systems	QL
55			In34 Performance of Municipal systems	i3	Reliability of municipal systems	QL
56				i4	Flexibility to meet technical innovation in municipal systems	QL
57	Building		In35 Certified green building (domestic or public)	b1	Requirement of certified green building (domestic or public) in blocks in regulatory plan	QL

4.3 Establishment of the final framework of assessment items

4.3.1 Introduction

As the research mentioned before, after the research selected and created a comprehensive initial framework of categories, indicators and parameters, a questionnaire will be made in order to establish the final framework of the assessment items. Through the questionnaire, the respondents will be asked to make suggestions to add new assessment items which they think are important and delete assessment items which are not considered to be important for the assessment or impossible for the practice by them. In order to make the result of the questionnaire more accurate, unitary and reliable, Delphi method will be employed in this part of research to collect and analyze all the data and to establish the final framework of the assessment items.

The Delphi method is a structured communication technique, originally developed as a systematic, interactive decision-making by a group of experts. (Harold A. Linstone, Murray Turoff, 1975) The method's name "Delphi" comes from the ancient Greece Oracle of Delphi. This method is established based on the assumption that group judgments are generally better than individual judgments.

The Delphi method was developed at the beginning of the Cold War to forecast the impact of technology on warfare. Then in 1950's and 1960's, it was developed by Olaf Helmer, Norman Dalkey, and Nicholas Rescher of Project RAND. Even since then, because of its accuracy and practicability the Delphi method has been widely used, together with various modifications and reformulations.

Generally speaking, there are six steps of the Delphi method procedure. For the first step, the problem is defined. Secondly, the panel members are selected based on their expertise. Thirdly, the questionnaire is prepared and disturbed to the panel members. For the fourth step, the questionnaire responses are collected and analyzed. Fiftly, the results based on the analysis of the last step are returned feedback to the panel members and see whether they can have a consensus. The fourth and fifth steps maybe repeated several times until a consensus is reached by the panel members. At last, the researchers make the final report based on the final results of the questionnaire.

According to the characters of this research, a new flowchart based on the standard flowchart of Delphi method is designed and adopted in the first part of questionnaire.

It is can be seen from the flowchart that questionnaire is the core part of the Delphi Methodology. After studying several reference cases, the questionnaire of this research is designed to comprise two parts which are Questionnaire 1 and Questionnaire 2, together with relational assistant introduction and explanation. Questionnaire 1 is used to collect the information of the initial assessment parameters from the two groups of respondents in three aspects which are understandability, utility and practicability. These three aspects are all essential for judging whether an assessment parameter can be adopted by the assessment tool and then be applied in the assessment practices. Questionnaire 2 is designed to collect the information from the two groups of respondents on which assessment items should be adopted by the final framework. An introduction of the initial framework of assessment items will be given to all the two groups of respondents first. And then, they will be asked to delete assessment items which they think are not important for the assessment or infeasible for the practice and make suggestions to add new assessment items which they think are important and feasible for the sustainable residential district assessment but not mentioned in the initial framework.

4.3.2 Sampling procedure and Questionnaire 1& 2

Respondents Selection

The respondents of questionnaire include a collection of stakeholders from different fields: urban planning and architectural design, engineering, construction management offices, estate developers, students, etc. According to their knowledge on sustainable development and assessment, they are divided into two groups: The first group is composed of experts and professionals from the urban planning and architectural design institutes, academic institutes, government offices of construction management and estate development companies. They are all involved in the urban residential district and eco-city development in Tianjin (Tientsin) of China. The list of experts and professionals is conducted with the help of Tianjin University and Tianjin Urban Planning & Design Institute, in which the researcher had studied for 8 years and worked for 4 years. The second group is composed of laymen who are undergraduate students concerning on the sustainability issue. Now the public participant is considered to play an important role in sustainable development and sustainability impact assessment in EU. And in developing countries, the public also has strong influence on the sustainable development and green building practices (Hikmat et al, 2009). So the non-professional's respondents of this questionnaire can be thought as a measure to encourage more public participants in China.

The respondent number of each group is defined according to their role and influence on sustainable development practices. The total number of respondents was 60: 50% experts and 50% laymen (non-professionals).

Structure of Questionnaire

Questionnaires are research instruments composed of a series of questions and other hints. They are often used to collect information and make statistical analysis from the respondents. The questions of the questionnaires can be divided into two types: open and closed. Open questions need the respondents to formulate their own answer while the closed questions need them to choose from several prepared options. And the options of the latter type should be exhaustive and mutually exclusive. Generally speaking, the closed questions are much easier to evaluate and elaborate results and very useful to quantify the information and make statistical analysis.

Considering the characters and needs of the research, a questionnaire composed of a series of open and closed questions is adopted in this part of the research. Closed questions are designed to gather information on the initial framework of assessment items prepared in the previous step of the research from the respondents. They can judge the performance of the assessment parameters according to three aspects- Understandability, Utility and Practicability. Besides that, they are also asked to delete those assessment items that they think are not important for the assessment and infeasible for the practices. The open questions are used to add necessary assessment items in case of the omissions of initial framework. And in order to make the respondents understand objects and process of the questionnaire better, the assessment items and questionnaire are introduced before the questions.

The submission way of the questionnaire is also important for the accuracy and efficiency. In such case, electronic data is adopted as the submission way for the experts group and the non-professional group because the questionnaire includes numerous questions which would take a long time to answer. All the respondents are asked to answer the questionnaire independently, which is required by the Delphi method in order to reduce the influence of conformism.

Then the questionnaire is designed to two parts. The first part is the introduction of the research and the questionnaire together with the description of initial framework of assessment items to make sure that the respondents of the questionnaire can make a well understanding of the research, questionnaire and assessment items. Because the assessment parameters represent special action directly, this level is selected as the core level of the questionnaire which is explained in detail. The second part is tables of Questionnaire 1 and 2 with closed and open questions for selection and suggestion. In Questionnaire 1, all the respondents will be asked to assess all the assessment parameters performance in three aspects. Then in Questionnaire 2, the respondents are asked to delete at least 6 assessment items (about 10% of the total number) which are not important for the assessment or infeasible for the practice and make suggestions to add new assessment items which they think are important. Then, all the answers of questionnaire respondents are collected and analyzed, and the results will be tabulated and returned to the participants of the questionnaire to see whether they can make an agreement on it or still have different opinions. If they cannot make a consensus, the questionnaire will be prepared and distributed again. All the procedure will be repeated until the participants can make a consensus. After that, based on the result of the questionnaire, the final framework of assessment items will be established.

The questionnaire is shown as Fig 4.2, Table 4.7, Table 4.8, Table 4.9 and Table 4.10:

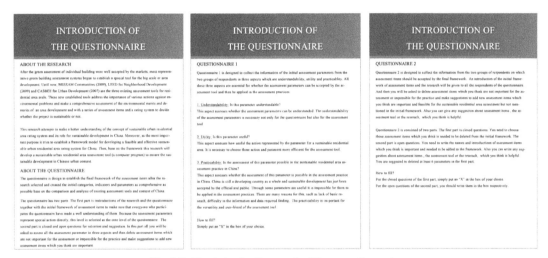

Fig 4.2 The introduction part of the questionnaire

Table 4.7 Example of the brief introduction of the initial assessment parameters

N°	Initial indicators		Initial parameters	Description	Requested data
	Land & Site					
1	In 1	Site selection	ls1	Retrofit area of land used	To assess how much area the development occupied comes from land used retrofitting. Its purpose is to protect the arable land, forest and other unused land.	Retrofit area of land used Total construction land area The document of regulatory plan
2			ls2	Impact of the development on the landscape	Analysis and protection of the landscape context's major features, of the land's nature and of the current views. Analysis and protection of the landscape context's major features, of the land's nature and of the current views.	The document of urban design and regulatory plan
...	...					
	Building					

| 57 | In35 | Certified green building (domestic or public) | b1 | Requirement of certified green building(domestic or public) in blocks in regulatory plan | To assess the requirement of certified green building (domestic or public) in blocks in regulatory plan. | The requirement in regulatory plan |

Table 4.8 Example of the detailed explanation of initial assessment parameters

	Land & site	
In1	Indicator	Site selection
ls1		Retrofit area of land used
	Typology	Quantitative
		Assessment Methodology
Assessment		R_r = Ratio of retrofit area of land used [%] $$R_r = \frac{A_r}{A_{tot}} \cdot 100\%$$ A_r = Retrofit area of land used [m²] A_{tot} = Total construction area [m²]
Method		
Points	0	20% > R_r
	1	50% > R_r ≥ 20%
	2	80% > R_r ≥ 50%
	3	100% > R_r ≥ 80%
	4	R_r = 100%
	Similar assessment items in existing systems
	New
	Reference laws and regulations
	China: Land administration law of PRC Code for classification of urban land use and planning standards of development land GB 50137-2011

Table 4.9 Example of the table of Questionnaire 1

	Initial Assessment Items		Understandability			Utility			Practicability		
N°	Indicator		Difficult	Medium	Easy	Inessential	Medium	Vital	Difficult	Medium	Easy
			Land & site								
1	In1 Site selection	ls1 Retrofit area of land used									
2		ls2 Impact of the development on the landscape									
3	In2 Compact development	ls3 The density of residential development									
4		ls4 Site coverage									
5	In3 Soil conservation	ls5 Environmental recovery of the excavated soil									
6		ls6 Soil permeability									
	...										
			Building								
57	In35 Certified green building (domestic or public)	b1 Requirement of certified green building(domestic or public) in blocks in regulatory plan									

Table 4.10 Example of the table of Questionnaire 2

Nº	Indicator	Parameter	Initial Assessment Items — Parameters need to be deleted from the final framework	New parameters need to be added	New indicators need to be added	New categories need to be added
		Land & site				
1	In1 Site selection	ls1 Retrofit area of land used				
2		ls2 Impact of the development on the landscape				
3	In2 Compact development	ls3 The density of residential development				
4		ls4 Site coverage				
5	In3 Soil conservation	ls5 Environmental recovery of the excavated soil				
6		ls6 Soil permeability				
		Energy				
		Water				
		Materials				
		Waste				
		Ecology				
		Transport				
		Outdoor environment & climate				
		Community economy & sociality				
		Infrastructure				
		Building				
57	In35 Certified green building (domestic or public)	b1 Requirement of certified green building(domestic or public) in blocks in regulatory plan				

4.3.3 Result analysis of Questionnaire 1 & 2

The analysis process of the questionnaire results can be divided into four steps, which are:

1. Analyzing the results of Questionnaire 2 of the experts and laymen groups, deleting those assessment items which are selected as the final assessment items by less than 50% respondents of the two groups and adding assessment items which are suggested to add by more than 50% respondents.

2. Analyzing the results of Questionnaire 1 to find the reasons why the respondents add or delete these assessments item, and then see which part of the initial assessment items need to be improved.

3. Establishing a new framework of assessment items and sending it to the respondents to see whether they still have different opinions or new suggestions.

4. Establishing the framework of final assessment items when the respondents can make a consensus.

Then after the questionnaire, as the first step the Questionnaire 2 results of all respondents are selected and analyzed. There are 3 assessment parameters selected by more than 50% respondents as parameters needing to be deleted from the final framework, which covered 5.3% of the total initial assessment parameters. The 3 assessment parameters are:

ls4 Site coverage

en5 Thermal efficiency level

en6 Electrical efficiency level

Thus, the left 54 of 57 initial assessment parameters which are supported by more than 50% of the respondents are selected into the new framework of assessment items that would be made into a new questionnaire to see whether they still have different opinions or new suggestions.

Meanwhile, there were no additional assessment items suggested by many respondents in Questionnaire 2, which proves that the initial framework of assessment items is comprehensive and integrated. 2 Experts thought that sun-shading of the window should be added into the assessment tools but they didn't give more information and method of this parameter and finally the suggestion was not accepted. Besides that, there were no other suggestions either on adding new assessment items and improving the detail of existing items.

Then after a further analysis and study of Questionnaire 1 of the respondents, the reason why they didn't choose these parameters is found.

The first step of the analysis is on the understandability of the initial assessment items.

For the Experts Group, because of their knowledge and experience on the sustainability issue, few respondents found difficulty in the understandability aspect. It is believed that the understandability would not be a problem both for the experts in the questionnaire and for those who take participant in the assessment practice in future. The university students of the laymen group have less knowledge and experience on the sustainability issues and projects operation than the experts. Even though, majority of them selected medium or easy on the understandability aspect of most assessment parameters. It indicated that with the detailed instruction and explanation of the parameters, the difficulty of understanding most parameters is accepted both for the respondents and for the users of the assessment tool.

Then the results on the utility and practicability are analyzed. Among these 3 unselected parameters, 2 of them were considered important but not selected by most respondents because of their poor practicability, which are en6 Electrical efficiency level and en5 Thermal efficiency level. The parameter of ls4 Site coverage was considered not important by some respondents and difficult in the assessment practice by some others. There were no assessment items selected as parameters need to be deleted from the final framework by the respondents due to their poor importance.

A further interview to some experts who didn't support these three parameters shows the special reason for each parameter. For the Site coverage, they thought the parameter of Density of residential development can measure the efficiency of land use very well and more practicable in the assessment. In such case, the parameter of Site coverage is not necessary for the assessment. For the thermal efficiency level, more and more municipal governments pay much attention on the thermal efficiency of buildings and publish much higher local standards especially in the eastern part of China. Enhancing the thermal efficiency is considered as the most important measure to reduce the energy consumption of building sector, so it is also strictly required and inspected during the governments' permit process of construction engineering. Another problem of this assessment parameter lies in the standard of climatic region in China which defines seven climatic regions with different requirements of thermal standard and heating system. Several climatic regions are not required to provide heating system because of their relative warm climate in the winter. However, in order to enhance the thermal comfort in winter many households of some part of these climatic regions use air conditioners for heating. Such a low thermal efficient way urges a better national or local standard but until now there have been no documents to publish. As a result, now it is much difficult in evaluating the thermal efficiency with a sustainable residential district assessment tool. Maybe in China the existing method which requires and inspects the thermal efficiency independently and locally is suitable for the status quo and practice better. For the Electrical efficiency level, many experts argued that though high electrical efficient measure will reduce the electrical load, it is impossible to reduce the capability of electricity supply which is required in the national standard.

Generally speaking, the results of Questionnaire 1 & 2 indicated that the initial framework of assessment items is considered as comprehensive and integrated by the respondents.

4.3.4 Final framework of assessment items

Following the analysis of the questionnaire results of Experts Group and Laymen Group, counting is a necessary step to establish an improved framework of assessment items, which will be sent to the respondents of the two groups again to see whether they still have different opinions or new suggestions.

After counting the results of Questionnaire 2 of the Experts and Laymen Groups in the last step, 3 assessment parameters were deleted, which are:

ls4 Site coverage

en5 Thermal efficiency level

en6 Electrical efficiency level

Therefore, these 3 parameters will be not adopted by the new framework which is composed of 11 assessment categories, 34 assessment indicators and 54 assessment parameters. The new framework is shown as Table 4.11.

Table 4.11 The new framework of assessment items

N°	Assessment Items		
	Categories	Indicators	Parameters
1	Land & site	In1 Site selection	ls1 Retrofit area of land used
2			ls2 Impact of the development on the landscape
3		In2 Compact development	ls3 The density of residential development
4		In3 Soil conservation	ls4 Environmental recovery of the excavated soil
5			ls5 Soil permeability
6	Energy	In4 Usage of sunlight	en1 Mitigation of sunlight obstruction requirement

		Assessment Items	
7		In5 Renewable energy	en2 Energy efficiency in the public space
8			en3 Effective results of renewable energy
9		In6 Energy-saving during construction	en4 Evaluate energy-saving activities at the construction stage
10	Water	In7 Conservation of water bodies	w1 Conservation area of water bodies
11			w2 Consideration of water quality
12		In8 Conservation of aquifers	w3 Groundwater conservation
13		In9 Water use	w4 Usage of unconventional water resources
14		In10 Water-saving infrastructure in public space	w5 Water consumption saving of the infrastructures
15	Materials	In11 Locally sourced materials in public space	m1 Usage of locally sourced materials in public space
16		In12 Recycled and waste resourced materials in public space	m2 Usage of recyclable materials in public space
17			m3 Usage of materials made from waste in public space
18	Solid waste & management	In13 Waste equipment plan	sw1 Waste collection equipments plan in public space
19			sw2 Requirements of waste collection equipments plan in blocks
20		In14 Waste management	sw3 Classification, treatment and disposal of waste
21		In15 Reduction of by-products of construction	sw4 Reducing the generation of construction by-products
22			sw5 Sorting and recycling of construction by-products
23	Ecology	In16 Green land	ec1 Percentage of green land
24		In17 Conserving natural resources	ec2 Conservation of green spaces existing
25			ec3 Consideration of natural flora in the surroundings
26		In18 Creating ecosystem networks	ec4 Eco mosaic fragmentation
27	Transport	In19 Capacity of transportation systems	t1 Sufficient capacity of transportation systems
28		In20 Accessibility of public transport	t2 Accessibility of public transport

		Assessment Items	
29		In21 Locations with reduced automobile dependence	t3 Convenience of commercial concentrations
30			t4 Convenience of financial institution (including ATM)
31			t5 Convenience of administrative offices
32			t6 Convenience of medical facility
33		In22 Pedestrian traffic	t7 Securing safety in pedestrian area
34		In23 Green travel measures	t8 Existence of green travel measures
35	Outdoor environment & climate	In24 Mitigation of heat island	o1 Horizontal shaded area ratio
36			o2 Open space ratio
37		In25 Improving acoustic environment	o3 Measures to reduce the impact of noise
38		In26 Improving wind environment	o4 Measures to improving wind environment
39		In27 Improving outdoor public space environment	o5 Consideration for the positioning of heat exhaust
40			o6 Vertical planting
41	Community economy & sociality	In28 Sustainable community residents structure	c1 Proportion of affordable residence for low income families and residence with low rent
42		In29 Sustainable community security	c2 Universal design and logo
43			c3 Measures of crime prevention
44			c4 Securing open space as wide area shelter and proper evacuation routes
45		In30 Sustainable community culture	c5 Formation of local centers & fostering of vitality & communication
46			c6 Community involvement
47			c7 Conservation and use of historical and cultural assets
48		In31 Sustainable community management	c8 Acquisition of ISO 14001 certification
49			c9 Formation of urban context and scenery
50	Infrastructure	In32 Performance of Information systems	i1 Reliability of information systems
51			i2 Flexibility to meet changing demand and technical innovation in information systems

		Assessment Items	
52		In33 Performance of Municipal systems	i3 Reliability of municipal systems
53			i4 Flexibility to meet technical innovation in municipal systems
54	Building	In34 Certified green building (domestic or public)	b1 Requirement of certified green building(domestic or public) in blocks in regulatory plan

Then the questionnaire of new framework of assessment items is sent to the two groups of respondents again to see whether they still have any different opinions. The feedback indicates that the new framework is considered comprehensive, practicable and appropriate for China and both the Experts Group and Laymen Group have no different opinions or new suggestions on it.

4.4 Weighting coefficient system set-up

4.4.1 Introduction

The Sustainable Residential district Assessment Tool is a multi-dimensional method which respects different aspects. Therefore, developing a weighting system of assessment items is necessary and it is the following step after the assessment items have been determined. Such a weighting system can define the importance of each assessment categories, indicators and parameters on a sustainable residential district according to the context of China. With the weighting coefficient system, the assessment items of each level are weighted. The scores for each assessment item are multiplied by the weighting coefficient and then aggregated into summation. In fact, the sustainable residential district assessment is a kind of multi-criteria analysis which takes into account all the items and values involved in a decision making process and the establishment of weighting system is a kind of multi-criteria evaluation problem. In order to solve this problem, many decision techniques are developed in this area. In this research, we are going to apply the methodology of Analytic Hierarchy Process (AHP) developed by Saaty to determine the weightings of all the assessment items according to the participants' questionnaires results.

4.4.2 The methodology of Analytic Hierarchy Process (AHP)

In this research, the weighting coefficient system is determined by using multi-criteria analysis within which a very remarkable role is played by the methodology of Analytic Hierarchy Process (AHP). The Analytic Hierarchy Process (AHP) is a structured technique for organizing and analyzing complex decisions. Based on the theory of mathematics and psychology, it was developed by Thomas L. Saaty in the 1970s and has been extensively studied and refined since then (Wikipedia).

AHP method has been widely applied in a variety of decision situations, especially in group decision making in politics, business, industry, military, etc, because it can transform human subjective judgments into quantitative analysis based on the principles of decomposition, comparative judgments, and synthesis of priorities. Compared with those decision making techniques which are intent on getting a correct decision, the AHP method is always used to help the decision makers not only find an alternative which satisfies their goal but also understand the problems. It provides a method which could decompose a complex problem into a rational and comprehensive structure. It has been found that the AHP method is widely applied in multi-criteria decision making in planning and resource allocation and conflict resolution (Vargas, 1990). As emphasized by Saaty (1980), AHP method permits to integrate both quantitative and qualitative aspects of decision-making, which makes it an efficient and effective method when faced with complex problems.

Generally speaking, there are six steps of Analytic Hierarchy Process (Xu, 2012). The details of each step are described as follow:

STEP1: In the first step of the process the user defines the goal or objective of the decision process and then decomposes the goal into a multilevel hierarchic structure from the top goal through intermediate levels of criteria to the lowest level of decision alternatives.

STEP2: After establishing the hierarchic structure, comparative judgments are required for pair-wise comparisons of its various elements according to their impact or importance on the elements above their level in the hierarchic structure. The fundamental scale of Saaty is applied here to indicate the relative importance or impact with respect to a criterion over another in a paired comparison. This 1-9 scale in the Analytical Hierarchy Process (AHP) was first introduced by Saaty in the AHP decision making theory in 1970's (Saaty, 1980). In 1970's and 1980's, Saaty tested the 1-9 scale, the index scale and about 20 other scales in order to find out a suitable ratio scale for the pairwise comparisons of the AHP method (Saaty, 1980; Saaty, 1994; Saaty, 1996). After many tests, the 1-9 scale was finally applied in the AHP method while other ratio scales were discarded.

After the pair-wise comparisons, the results constitute a positive reciprocal matrix and the values of its elements are between 1/9 and 9. The diagonal elements of the matrix are equal to 1. The a_{ij} is equal to the value of comparison between element i and element j with respect to the considered criterion and the a_{ji} is equal to the inverse of the a_{ij}.

STEP3: After constituting the pair-wise comparison matrix, the consistency of the judgments will be measured by the principal eigenvalue of the matrices. It is a necessary step for deriving the priorities from the pair-wise comparison matrix. The principal eigenvalue should be 10% or less to prove that the consistency of the judgments can be accepted, otherwise the judgments should be made again.

STEP4: Hierarchic composition is used to weight the eigenvectors in a level by the eigenvector weights of the corresponding criteria and the sum is then taken over all weighted eigenvector entries in the next lower level of the hierarchy. The resulting priorities are thus determined with respect to the overall goal of the hierarchy.

STEP5: The consistency of the entire hierarchy is determined by multiplying each consistency index by the priority of the corresponding criterion and adding. The result is then divided by the same type of expression using the random consistency index corresponding to the dimension of each matrix weighted by the priorities of the corresponding criterion.

STEP 6: The best alternative is obtained by rating the hierarchies on single or group criteria and computing the priorities of each alternative by weighting and adding for the different levels.

Because of the large number of assessment items needed to compare, the software of Expert Choice is employed by this research to constitute the matrix and calculate the results. Expert Choice is collaboration software which is used to help groups make better decisions in multi-criteria decision-making and achieve agreements with a speedy, transparent and simple decision process. It is developed based on the theory of AHP methodology and engages decision makers in structuring a decision into smaller parts, proceeding from the goal to criteria to sub-criteria down to the alternative courses of action. Decision makers then only make simple pair-wise comparison judgments to derive overall priorities for the alternatives. In addition, Expert Choice can calculate the principal eigenvalue and measure the consistency of the judgments, which are extremely important for AHP method. Now this software is widely accepted in social, political, technical, and economic fields.

4.4.3 Developing the hierarchic structure of weighting coefficient system and Questionnaire 3

Developing the model of weighting coefficient system is the necessary step of applying the AHP methodology. As mentioned before, the user should firstly define the goal or objective of the decision process and then decomposes the goal into a multilevel hierarchic structure from the top goal through intermediate levels of criteria to the lowest level of decision alternatives.

In this research, the top goal is defined as "A Sustainable Residential District". Then the categories and indicators are defined as criteria and sub-level criteria. And finally, the parameters are defined as the alternatives which are the lowest level of the hierarchy.

Then seven respondents of Experts Group are invited to answer the Questionnaire 3 by making simple pair-wise comparison of different levels of assessment items. The theory and process of AHP methodology are introduced to them in order to make them understand the method better and reduce the inconsistency of their answers. And then, based on their knowledge about the assessment of sustainable residential districts, using the AHP method, they make the pair-wise comparison of assessment items. Then the weights of all assessment items are determined by analyzing the judgments of the experts.

It is well known that testing the validity of AHP models developed in the research is an important analysis. The widely accepted validation of eigenvector method for estimating weights in the AHP yields a way of measuring the consistency of the experts' preferences arranged in the comparison matrix. Furthermore, this validation method mainly refers to the consistency ratio (CR) parameter of the pair-wise comparison matrices. Thus the consistency ratio (CR), indicating the probability that the matrix judgments are randomly generated, is set to be less than 0.1. Otherwise it is necessary to re-evaluate or revise the matrix judgments. Here, details of the two formulas are defined as following.

Where CR is the consistency ratio, CI is the consistency index, λ_{max} is the principal eigenvalue of the matrix, and n is the order of the matrix. RI is the calculated Random Inconsistency Index for the matrix given an order n (Saaty, 1977). All the matrices of each level should be measured in terms of the consistency, and the CR of them should be less than 0.1.

4.4.4 Result and analysis

All the answers of Questionnaire 3 are collected and then input into the software of Expert Choice and using AHP methodology to calculate the weighting coefficient of each assessment items. Based on the function of combining all the participant experts' choice, an overall pair-comparison of assessment categories with respect to the goal— A Sustainable Residential District are calculated by the Expert Choice software Fig 4.3.

	Land & site	Energy	Water	Materials	Solid waste	Ecology	Transport	Outdoor en	Community	Infrastruct.	Building
Land & site		1.48599	1.48599	2.11927	1.73851	1.10409	2.80021	1.84218	2.37957	1.84218	2.11927
Energy			1.10409	2.18791	1.98165	1.04195	3.3064	1.79482	2.39348	1.42616	1.42616
Water				1.21901	1.21901	1.29171	4.36599	1.66851	1.99551	1.10409	1.10409
Materials					1.0	1.91947	5.13452	2.41565	4.60777	1.10409	1.21901
Solid waste & management						1.73851	5.40292	3.3064	5.02268	1.04195	1.36874
Ecology							2.24565	1.73851	2.77275	1.29171	1.42616
Transport								1.64067	1.21901	2.71235	2.04583
Outdoor environment & climate									1.29171	1.57461	1.57461
Community economy & sociality										2.35355	2.41565
Infrastructure											1.21901
Building	Incon: 0.01										

Fig 4.3 The combined result of pair-wise comparison of assessment categories with respect to the goal- A Sustainable Residential District

Then all the pair-comparisons of assessment indicators and parameters with respect to the criteria of each level are calculated.

As mentioned before, in order to make sure that the judgments made by the experts are acceptable, the pair-wise comparisons matrix needs to be consistent. The consistency ratio (CR) is used to measure the consistency of matrix and if CR is less than 0.1 the judgments are accepted (Saaty,1980). As an important function, the Expert Choice software can calculate the CR automatically both for the expert's individual pair-wise comparisons and for the combined results. And after the calculation, all the CR is less than 0.1 and the judgments of experts are accepted.

Then based on the comparisons of matrices combining all the experts' individual judgments, all the weights of the assessment items are calculated by the Expert Choice software. Then the weights of assessment parameters are also calculated by the Expert Choice software.

4.4.5 Comparison among the categories' weighting coefficient of BREEAM Communities, LEED for Neighborhood Development, CASBEE for Urban Development and SRD rating system

Based on the preliminary research and comparison between the assessment items of BREEAM Communities, LEED for Neighborhood Development, CASBEE for Urban Development, common categories are defined and also used as the final assessment categories of SRD. After establishing the weighting coefficient system, it is necessary to make a comparison of the categories' weighting among BREEAM Communities, LEED for Neighborhood Development, CASBEE for Urban Development and SRD Rating System.

Then, the assessment items' weighting of these tools is counted again in terms of SRD assessment categories. Fig 4.4 indicates the percentage that each category contributes in different tools.

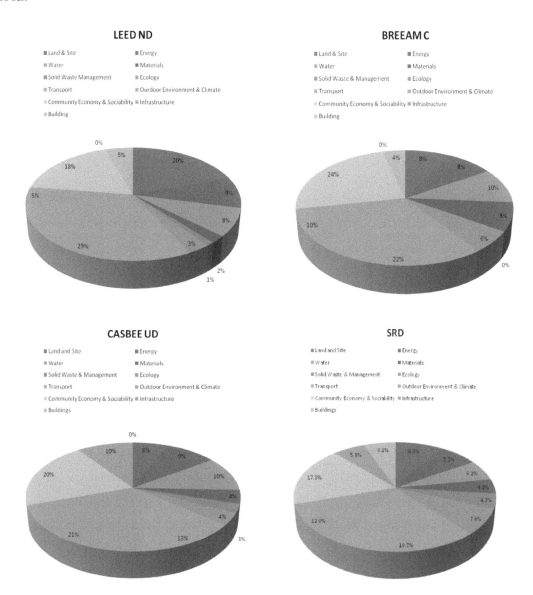

Fig 4.4 Comparison among LEED ND, BREEAM C, CASBEE ND and SRD in terms of SRD criteria assessment categories

From the comparison and analysis, it is noticed that transport and community economy & sociality are considered as the most two important categories which contribute 19.7% and 17.8% respectively in SRD in China. It is very similar with the weighting systems of the other three tools. In LEED for Neighborhood Development, the weights of transport and community economy & Sociality are 29% and 18%, rank first and third in all the categories. In BREEAM for Community, their weights are 22% and 24%, rank second and first. In CASBEE for Urban Development, their weights are 13% and 20%, also rank third and second. Another important category in SRD is outdoor environment & climate which takes 12%. Such weight is similar in BREEAM C (10%) and CASBEE UD (21%). Compared with LEED ND (5%), the higher residence density is considered as the reason for this.

4.5 Sustainable Residential District (SRD) Rating System for China

4.5.1 Introduction

A user-friendly tool is necessary for the success of sustainable residential district assessment because it can make the assessment practice more convenient and easier to be accepted by the urban planners, development companies, government management institutes, expert assessors, etc. In addition, such a tool would provide more information of the assessment projects with which the participants can understand the different aspects of projects' sustainability issues better and make useful comparison and analysis within or between projects.

So as the next step of the research, an assessment system should be developed based on the previous research that included the final assessment categories, indicators, and parameters and their weighting coefficient. In this research, this assessment system is defined as a criteria-based tool which assigns credits to a series of items on a certain scale ranging among five levels—Excellent, Good, Satisfactory, Pass and Unclassified. The middle level-Satisfactory is the average performance or the national standards. This system is designed to implement during the interim (optional) stage and final (mandatory) stage, which makes it could cooperate with the residential district development process and national evaluation standard for green building of China. Besides that, now the assessment system is only designed for new developments of residential districts. The special edition for assessing regeneration projects will be developed in future after the edition for new developments gets mature.

Then, a computer program was created to calculate the overall performance level of sustainable residential district. The excel spread sheet based system consists of the main sheet that includes basic information of the residential district projects, results sheet that includes the overall assessment results such as the ranking level and the radar chart of categories' score, the detail assessment results of each category, the score sheet which can be considered as a score and weight data checklist of all the assessment parameters, indicators and categories and 11 detail assessment sheets of the following categories: Energy, Land Use &Site, Water, Materials, Solid Waste & Management, Ecology, Transport, Outdoor Environment & Climates, Community Economy & Sociality, Infrastructure, Buildings. Each sheet of categories includes the assessment indicators, parameters, brief introduction of their assessment method, the 0−4 points score criteria and the score choice with drop-down list.

In order to make analysis and comparison of the assessment projects, the score of each category or indicator is calculated independently. It comes from counting the parameter's points multiplied by the specific parameter's weight. The score of each parameter will appear on the indicator level and then the score of indicator level will be multiplied by their relative weights and also be shown on the category level. Finally, the score of each category will be multiplied by their weight and be show on the overall level. All these values will be shown on the main result sheet and indicate the contribution of each category and be presented graphically to show the overall sustainable performance level of the assessment residential districts.

It should be pointed out that in SRD Rating system the ranking benchmark is based on the overall score gained percentage. The final ranking of the performance of the assessment projects depends on the score percentage of summation above or below a given number. According to the analysis of the existing rating systems such as LEED and BREEAM, five performance levels are designed, which includes Excellent (≥ 85%), Good (70%−85%), Satisfactory (50%−70%), Pass (30%−50%) and Unclassified (<30%).

Another important part of the system is the technical handbook which consists of an introduction of the tool and detailed explanation of assessment parameters. Such a technical handbook can make all the participants of the projects especially the assessment operators understand the tool better. Besides that, in the interim stage the handbook can be used as a guideline by the decision makers such as the development companies or the urban planners and designers. Such an instrument can guide them to make better choices on the sustainability issues which would have significant positive influence on the development of sustainable residential district.

4.5.2 The excel spread sheets of Sustainable Residential District(SRD) Assessment Rating System for China

As mentioned before, the excel spread sheet of the SRD Rating System consists of four parts: the main sheet, the results sheet, the score sheet and the assessment sheets of all categories. As the first part, the main sheet consists of basic information of the assessment residential district project, such as project name, location and year of completion.

Then, the results sheet consists of three parts: summary of the designated area, overall score level and assessment indicators results of each category. In this sheet, the assessment results are illustrated by bar charts and radar charts with the purpose of comparing the scores between different indicators, categories and projects. Such a comparison can make the users know exactly which parts of the assessment residential district project need to improve and which way to improve is better.

The score sheet can be considered as a score and weight data checklist of all the assessment parameters, indicators and categories. All the score judgments of the users will be shown here, together with the weight of assessment items respectively. The sheet will calculate the score of each parameter, indicator and category and present them in this sheet.

Moreover, 11 sheets of the assessment categories are designed to simplify the assessment work. Each sheet includes the assessment indicators, parameters, introduction of their assessment methods, the 0–4 points score criteria and the score selection with drop-down list. The category sheets only provide basic information of the assessment parameters. In the assessment practice, the assessors should use it with the technical handbook which explains more details of the assessment parameters.

4.5.3 The technical handbook of Sustainable Residential District(SRD) Assessment Rating System for China

The technical handbook of Sustainable Residential District(SRD) Assessment Rating System for China consists of two parts: an introduction of the tool which is mainly used to help the users make a general understanding of the tool and an explanation of the assessment parameters' details which provides the users with more information about the parameters and teach how to make the assessment of a special parameter.

The brief introduction of SRD Rating System

Aim and role

The Sustainable Residential District (SRD) Assessment Rating System is designed to become a feasible and efficient assessment tool for Chinese residential district development on the issues of sustainable development. Such a tool can help the estate developers, urban planners and designers and government management institutes improve measures and certify the sustainability of developments at the residential district scale in the context of China.

The SRD Rating System is established based on the system of urban planning, estate development, construction management and government structure in China. Similar with other assessment systems such as BREEAM, CASBEE and LEED, it is designed to cooperate with the assessment tool for building scale— Evaluation Standard for Green Building (National Standard of the People's Republic of China) which now is the most widely accepted assessment tool in China.

Scale of assessment

Now in China, according to the National code for urban residential district planning & design, the residential development can be rated into three levels: residential district, residential quarter and neighborhood. They each have their respective scale, population and requirements of municipal services. The national code defines them as following:

Residential district refers to those residential settlements surrounded by arterial roads or natural boundaries and generally have a population of 30,000－50,000 or more.

Residential quarter refers to those residential settlements surrounded by urban streets or natural boundaries and generally have a population of 10,000－15,000.

Neighborhood refers to those residential settlements surrounded by residential quarter roads and generally have a population of 1,000－3,000.

The different scales correspond to different steps of residential development in China. Considering that, the residential district is selected as the assessment scale of the assessment system because such scale can make the tool cooperate with the regulatory detailed plan and the land exploiting of residential development better, and make it more useful and efficient in promoting the sustainable development of cities.

Assessment scope

Considering the characters of residential development in China, the scope of assessment is divided into two kinds of spaces as following:

1. The public space

2. The blocks

In the public space, all the construction or development are carried out by the land development companies (state-owned). All the measures against the sustainable objectives can be finally assessed by the construction practice with the tool.

In the blocks, all the measures against the sustainable objectives are assessed by the requirements in the approved regulatory plan with the tool. The measures will be carried out by the building development companies.

Assessment type

Now the assessment tool is designed for new developments of residential districts. The special edition for assessing regeneration projects will be developed in future after the edition for new developments gets mature.

Assessment stage

In order to cooperate with the assessment tool for building scale- Evaluation Standard for Green Building (National Standard of the People's Republic of China), the regulatory detailed planning system, the procedure of residential development, the construction management system and government structure, two assessment stages are defined as following:

1. The Interim Stage (Optional)

In the interim stage, the assessment tool is used to measure the commitments against the sustainable objectives and planning policies by the initial plan of the public space and blocks. The certificate of this stage is not a final assessment rating. It is used to test the planning and forecast future performance of the assessment residential district project. It is an optional stage of assessment. Meanwhile, the tool can also be used as a guideline for the planning designing or decision making in this stage.

2. The Final Stage (Mandatory)

In the final stage, the assessment tool is used to measure the commitments against the sustainable objectives and planning policies by the approved regulatory planning and the completed construction in the public space. The certificate of this stage is a final assessment rating.

Target users

The SRD Rating System is designed to be used by the estate developers, the urban planners, the designers, the government institutes, the expert assessors, etc.

Assessment items

The framework of the SRD Rating System assessment items consists of three levels: 11 Categories, 34 Indicators and 54 Parameters. More information of each special assessment item will be provided in the latter detailed explanation.

Chapter Five: Tests of the Sustainable Residential District (SRD) Rating System for China

第5章 我国大型城镇住区可持续性评估工具的应用检验

为了验证可持续居住区评估工具的实用性与可操作性，研究选取了位于天津市的住区项目作为评估样本。在这一章中，将对评估过程中出现的问题及所得结果进行深入分析与对比，以期总结经验教训，并为后续研究提供有益的参考与建议。为确保评估工作的全面性与专业性，研究邀请了本地专家全程参与，从专业使用者的角度出发，为完善可持续居住区评估工具提供宝贵的建议，从而推动其在实际应用中的不断优化与发展。

首先，对 4 个测试案例的评估结果进行深入比较与分析，不仅有助于改进评估工具，更能深化对这些项目可持续性问题的理解。这 4 个测试案例均代表了可持续性问题的平均性能水平，具有一定的代表性。经过严格的评估实践，其中两个测试案例获得了"满意"的评级，一个获得了"良好"的评级，另一个也达到了"通过"的标准。尽管这 4 个测试案例在总体性能上存在一定的差异，但综合考量其表现，可以认为评估结果是合理且可信的。

总体而言，土地在场地、水资源以及生态三大类别中表现尤为突出，平均得分逾 70%，即 2.8 分。这一评估结果的成因颇为复杂。就土地与场地而言，近年来，中国政府、开发商及城市规划者已广泛接纳新城市主义与紧凑发展的理念。因此，在规划与开发的实践中，更加注重土地使用的优化、开发密度的提升以及对景观影响的最小化。在水资源方面，现存水体通常被视为宝贵的景观资源，与绿地相结合，共同打造成公园或公共开放空间，旨在提升居住品质与建筑价值。故此，多数水体在开发过程中得到了妥善保护。同时，为维持水质，近期采用了净水处理、人工湿地等自然净化手段。政府、开发商及城市规划者亦高度重视水资源的节约利用。多数住宅区在开发中均建立了灰色供水系统，以分离生活污水，为景观与卫生设施提供再生水。此外，还颁布了国家标准以推广节水装置。至于生态学，往往被我国公众视为可持续发展的核心要素，绿地比例则被视为衡量生态状况的首要指标。尽管人们对这一观点存在误解，但却推动了中国住宅区开发中绿地规划的不断增加。

然而，相较于前述三个类别，其他类别的表现则显得逊色。在八个类别的多数评估参数上，项目的表现往往只能达到中等水平，得分为 2 分。由于缺乏更具针对性的指导方针，多数开发商、规划者及设计师在推进项目时，往往仅满足于达到国家标准的基本要求，或在可持续性问题上追求平均水平的达成。这种倾向可能限制了项目在可持续性方面的进一步发展和优化。

从测试评估案例的反馈结果来看，大型城镇可持续居住区（SRD）评级系统在实用性和有效性方面均展现出了超出预期的表现。开发商和规划者普遍认为，SRD 评级系统作为一种全面而清晰的工具，为他们深入理解和评估住区项目的可持续性问题提供了有力的支持。在实际应用中，该系统不仅操作简便，难度适中，而且其基于研究现有评估体系所构建的评级水平假设量表,在测试中也被证实具有良好的可接受性。

然而,尽管 SRD 评级系统总体表现良好,但仍存在一些需要改进的问题。由于部分评估参数直接借鉴自发达国家的现有评估工具,并首次应用于我国的评估实践中,因此在某些方面得分较低。为了进一步提升评级系统的适用性和准确性,未来在更多的评估实践中,需要根据我国本地的实际情况对这些参数进行细致的调整和优化。

5.1 Introduction

As mentioned at the beginning chapters, in China the residential development on the residential district scale is growing vigorously in recent years. Here, four residential district development projects are assessed by the suggested Sustainable Residential District (SRD) Rating System as a test to identify the practicability and usefulness of the tool. The problems found in the tests will be analyzed by the research to find more useful information and to give suggestion for future research.

All the four test projects are located in Tianjin which is a metropolis in northern China. They are selected as test objects because they all have suitable scales, and represent the average performances of Chinese residential district projects on the sustainable issues nowadays. Of course, all the test projects locating in a same place would bring problems and limitations. So in future, the research will test more projects from other places especially from different climatic regions of China in order to improve the applicability of the tool.

In the assessment process, Tianjin University and Tianjin Urban Planning & Design Institute offered much help. Several experts took part in the assessment work. They provided the required planning and designing files, data of the four test projects and help to connect with the local municipal governments, management institutes and companies .etc. Through the assessment, they thought SRD Rating System is a useful tool which allows them to make a general and much clearer view of the residential district projects. Such a tool can help them measure and compare different residential district projects on the sustainability issues. Besides that, by analyzing the assessment results provided by the tool they can find how to improve the performance of future residential district projects.

5.2 Test 1: The project of Huaming Town

Huaming Town is located in the northeast of Tianjin, closes to the airport of Tianjin Binhai International Airport. The Huaming Town Residential District project has a total area of 196.8 hm^2 and a planning population of 44,000 when project completed.

The project of Huaming Town started in 2005 and it is a new urban-rural integrated residential district in the course of Chinese urbanization. Huaming Town is also the first town built in the mode of "exchanging house site for house" which is considered as a successful model of peasants' happy life which is ecological, harmonious and habitable in Tianjin.

Ecosystem conservation is taken as the most important idea of the planning and designing. The "conserved ecosystem" reserves the ridges of fields and over ten thousands of trees on both sides. Based on that, the road system and green system of the residential area are built so that the main roads and buildings will be covered with tall trees and the raw type of surroundings can stand out.

Another planning idea is "maintaining inherent qualities and peace", which refers to organizing new yards between neighborhoods and forming a community of every four neighborhoods and a 3 levels structure of neighborhood community district with the public service facilities, public greenbelt and leisure square as the center and thus to provide residents with a convenient and comfortable residential environment.

Since the residents are peasants, the residential design should take the living habits of peasants into full consideration. Therefore, their representatives were involved in the planning, design, maintenance and management stages. Universal design is taken for the entrance of the buildings and public space. Besides that, the elevation designs adopt brief lines and plain materials to manifest the regional characteristics of a north town.

Energy-saving and release-reduced technologies are adopted in the building design. All the buildings adopt the constructional technology of "3-step energy conservation", such as heat preservation and heat insulation of outer wall, heat preservation system of insulated glass doors and windows, ventilated roof and heat preservation and insulation. Second, over 8,000 solar energy systems were installed and the integration of solar energy driven water heaters and buildings is realized. Third, the drainage of rain and sewage is reduced and the separate drainage of rain and sewage to retrieve is adopted. The sewage is used for landscape after being collected and processed through the artificial intensified processing units such as pretreatment system, biochemical system and sedimentation tank and finally constructed wetland so as to realize the circulation of water resources.

Huaming Town is considered as a successfully-built environment-friendly residential district which is ecological, harmonious and habitable in Tianjin when it was building. This project has been selected as the case of Urban Best Practices Area, 2010 Shanghai Expo. It is also considered as a successful example of rapid urbanization of Chinese rural area in China. The government thought it is typical and demonstrative, and can serve as a good reference of the urbanization for China and other developing nations.

The photographs of Huaming Town are shown in Fig 5.1.

The assessment results indicate that the overall sustainability performance of Huaming Town is on the satisfactory level. The scores of most assessment categories are over or near two points, which is the average sustainability performance level of the residential district development now. Considering the project, were planned and designed 8 years ago and had been completed for over 5 years, such results are accepted though the project was thought to be a model of sustainable residential district at that time.

After an analysis of the results, it is can be seen that there is much difference between the scores of various categories. Among the 11 categories, Land & site, Water

Fig 5.1 The photographs of Huaming Town

and Ecology are the three best performance categories with over 2.8 points which means they get an over 70% score percentage or good level rating. The categories of Transport, Outdoor environment & climate and Community economy & sociality get over 2 points or 50% score percentage. Energy, Infrastructure and Building are the three categories which get less than 2 points or 50% score percentage. The reason of poor performance of Energy category lies in the negligence of energy saving in the public space though much energy-saving technologies such as heat preservation and heat insulation of outer wall, heat preservation system of insulated glass doors and windows, ventilated roof and heat preservation and insulation, solar energy driven water heaters system adopted in building design in the blocks. When the projects are planned and designed, the sustainability issues of infrastructure didn't catch much attention, it is the reason of poor performance of Infrastructure in the assessment. When the project was built, the Evaluation Standard for Green Building (Standard by People's Republic of China, GB/T 50378−2006) has not been published. As a result, the Building category got 0 point in the assessment. But it doesn't mean the buildings in Huaming Town perform so poor in the sustainability issue. In fact, most experts participating in the assessment work thought that the building could get 2 stars level if an assessment were made according to the Evaluation Standard for Green Building. So 0 point is not an accurate result for the assessment, but it won't change the final overall score level of Huaming Town project.

5.3 Test2: The project of Balitai Town

Balitai Town is located in the southeast of Tianjin. The Balitai Town Residential District project has a total area of 383.49 hm^2 and a planning population of 48,000 when project completed. The project of Balitai Town also belongs to the program of "exchanging house site for house" in Tianjin.

The assessment results indicate that the overall sustainability performance of Balitai Town is on the Pass level. Similar with Huaming Town, the scores of most assessment categories are about two points, which is the national standard level or the average sustainability performance level of the residential district development. Among the 11 categories, Water and Ecology are the best two performance categories with a score of 2.653 and 2.710 respectively. Energy, Solid waste and management and Building are the three categories with the poor performance.

5.4 Test3: The project of Shuangkou Town

Shuangkou Town is located in the north of central area of Tianjin. The Shuangkou Town Residential District project has a total area of 377.16 hm^2 and a planning population of 113,000 when project completed. Different from the two projects assessed before, the FAR of Shuangkou

Town is much higher, reaching to 120%, due to the less distance from the city centre.

The assessment results indicate that the overall sustainability performance of Shuangkou Town is on the satisfactory level, but because the project is planned and designed more recently than the two projects assessed before, it got a better score percentage which is 58.23%. The best three performance categories are Land & site, Water and Ecology, which got a score of 2.866, 2.816 and 3.005 respectively. For the category of Building, the Evaluation Standard for Green Building (Standard by People's Republic of China, GB/T 50378-2006) has been adopted by the project, so it got a better score on this aspect.

5.5 Test4: The project of Gaojiazhuang Town

Gaojiazhuang Town is located in the north of Tianjin. The Gaojiazhuang Town Residential District project has a total area of 359.59 hm^2 and a planning population of 91,000 when project completed.

The assessment results indicate that the overall sustainability performance of Gaojiazhuang Town is on the good level. Because of the more recently planned and designed date the project got a much better score percentage than the projects assessed before, which is 71.04%. The best four performance categories are also Land & site, Water, Ecology and Building, which got a score of 3.000, 3.632, 3.300 and 3.000 respectively.

5.6 Comparison and analysis of the assessment results of the four test cases

A comparison and analysis of the assessment results of the four test cases is necessary for the research and the suggested rating system. It is helpful to improve the assessment tool and to make a better understanding of these projects on the sustainability issues.

As mentioned at the beginning of this chapter, all the four test cases represented the average performance level on the sustainability issues. Through the assessment practice, two of the test cases got a rating of Satisfactory, one got a Good level and one got a Pass level. Though there were some differences on the overall performance among the four test cases, the assessment results were generally considered as rational and acceptable.

Generally speaking, Land & site, Water and Ecology were considered as the three categories of best performance, which averagely got a more than 70% score percentage or 2.8 points. The reason of such assessment results lies in different aspects. For the category of Land & site, in China the conception and general ideas of New Urbanism and Compact Development have been accepted by the governments, developer and urban planners in recent years. As a result, the importance of retrofitting of land used, raising the density of development and reducing the impact of development on the landscape was recognized in planning and development practice. In the category of Water, the existing water body was generally treated as a landscape resource which was designed as a park or public open space with the green land to enhance the residential quality and the price of buildings so most water body was conserved very well in the developments. In addition, in order to keep the water quality, water purification treatment and natural purification such as the artificial wetland were adopted recently. Water use reduction and saving were also valued by the governments, developers and urban planners because the price of water is rather high in most parts of China. In most residential district developments, a gray-water system was built to retreat the sewage of living and to provide for the landscape use and the toilet flashed water. Special national standard was published to promote the water saving devices. For the category of Ecology, it was often treated as the most important aspect of sustainable development by the public in China and the percentage of green land was regarded as the most important indicator of the ecology. Such an opinion led to a trend that more and more green lands were planned in Chinese residential districts development though it was a misunderstanding. Besides that, with the development of Landscape Ecology in China, the conserving natural resources and creating ecosystem networks got more and more attention.

However, the performance of other categories was not so well as these three categories. On most assessment parameters of these 8 categories, the projects usually perform at the middle level and get 2-point score. Without more useful and efficient guides, most of the developers, planners and designers tend to make the projects just satisfy the requirements of the national standards or reach the average level on the sustainability issues.

From the feedback of the test assessment cases, the practicability and usefulness of the suggested tool— the Sustainable Residential District (SRD) Rating System is better than expected. The involved developers, planners thought the SRD Rating System is a very helpful tool which allows them to make a general and clearer view of the residential district projects on the sustainability issues. It is useful in the practice and the difficulty is acceptable. The hypothetical scale of the rating levels, which is established on the research of existing assessment system, is also considered to be acceptable in the test. Among the four cases, the newest planned project got the best score and Good level rating. The former model and a new project got a Satisfactory level and the relative old project got a Pass level. Such results represent their performance on the sustainability issues relatively and accurately. However, there are still some problems of the rating system. Because some assessment parameters of the rating system come from the existing assessment tool in the developed countries and are used in the assessment of China for the first time, the scores are very poor. Some details of these parameters are needed to adjust according to the local context, after more assessment practices.

Chapter Six: Conclusion and Future Direction of the Research
第6章 研究的结论和未来的研究方向

本章旨在对研究进行全面的归纳与总结，深入剖析住区的可持续性对于我国城镇化进程的重要意义，并基于研究过程中所揭示的问题与存在的局限性，对未来的研究方向进行合理的设想与展望。

本研究的核心旨在探索一种契合我国大型城镇住区发展的可持续性评估工具，以科学量化其在可持续发展问题上的实际表现。经过前几章的研究与实践应用，构建出可持续居住区（SRD）评级系统这一工具。通过对我国 4 个具有代表性的住区进行实证评估，验证了该工具的实用性和有效性。

在我国的住宅发展体系中，住区作为一个整体单元，在功能和结构上具有高度的统一性和完整性。因此，在这一层面上，可以实施诸多重要的街区规模可持续行动或措施，如提升公共交通的可达性、构建生态系统网络以及优化可持续的住区居民结构等。这些举措对于推动住区的可持续发展具有重要意义。

同时，控制性详细规划以其强大的实用性和可操作性为基础，使得基于其的评估工具能够有效地整合和控制方案、设计、施工以及管理过程。这种整合和控制有助于确保可持续目标的顺利实现，为住区的可持续发展提供有力保障。住区评估工具通过其综合性的考虑、评估和要求，能够为同一地区的不同建筑开发公司提供更为优质、公平的绿色建筑建设条件。这有助于减少企业之间的低价低性能竞争，推动绿色建筑行业的健康发展。住区规模和控制性规划平台的结合，使得评估工具能够与建筑规模评估工具《绿色建筑评估标准》更好地协同工作，而无须担心技术上的冲突或障碍。这种协同作用有助于提升评估工作的效率和准确性，为住区的可持续发展提供更为全面、科学的支持。

在构建可持续居住区（SRD）评级体系之后，运用该工具对 4 个代表性大型城镇住区进行了评估。这些住区均具备适宜的规模，且为近年来规划、开发与建设而成，因此能够有效地反映我国大型城镇住区在可持续发展方面的整体表现水平。从测试评估案例的反馈来看，评估参与者对所提议的 SRD 评级系统的实用性和有效性表示认可。相关开发商和规划师普遍认为，该系统为他们提供了一个较为全面的视角，以审视住区项目的可持续性问题。同时，该系统在实践中的可操作性也得到了验证，其难度在可接受范围内。基于现有评估体系研究所建立的评分水平假设量表同样被认为具有可接受性。对 4 个测试案例的评估结果进行比较分析，揭示出我国大型城镇住区在可持续发展方面存在的一些问题。总体而言，这些住区在土地利用、场地规划、水资源管理和生态保护方面的表现相对较好，但在其他方面仍有待进一步改进。

未来的相关研究首先应致力于收集与分析我国不同地区的更广泛的数据和项目。通过吸纳来自各地区和城市的专家参与研究，可以对建议的评级体系进行更为细致的调整，并针对特定地区或城市开发相应的子版本，从而提升其在当地背景下的实用性和适用性。

其次，应探索构建针对其他类型片区尺度的评估工具，这样的系统可以为城市的可持续发展提供更为全面、多元的指导。最后，将更多来自不同地区和城市的开发项目纳入建议的评级系统进行评估。通过对评估结果进行比较和分析，可以测试该系统在不同背景下的实用性，并识别出存在的不足之处以及需要改进和解决的问题，这将有助于我们不断完善和优化评级系统，推动其在实践中的应用和成熟。

6.1 Introduction

This chapter tries to summarize the main findings and results of this research and analyzes its contributions and benefits for the urbanization and residential development in China. Then it points out the limitation and deficiency of current research and the future direction.

6.2 Research Conclusion

The main purpose of this research is "exploring a sustainability assessment tool for big scale residential area development in China, to measure the performance on the sustainability issues, and thus to promote the sustainable development in the urbanization of China." This purpose has been successfully achieved through the previous chapters of this research. The outcome of this research established the suggested Sustainable Residential District (SRD) Rating System (a computer program). With this tool, four residential district developments in China were assessed and the result proved its practicability and usefulness.

Meanwhile, the derived objectives of the research were also achieved. Through the research, the state of sustainable development and green building in China were investigated. Now the urbanization and residential building construction is a key element for China to keep the rapid pace of the economic growth. China's rapid economic expansion is driven by the rapid urbanization, by the huge amount of land exchanged and by construction materials and equipment produced for using in buildings. In order to solve the conflicts between environmental pressure and rapid urbanization, sustainable development has become an important development strategy in China. In recent years, the research of green building and green building assessment has made a great progress. Several assessment methods of Green Building are produced. Among them, the Evaluation Standard for Green Building is a National Standard, which started to be used in some developed regions and cities in China. But until now, there have been no assessment tool for area scale development in China.

Different from those assessment tools for buildings, the scale and platform are essential for the assessment tool of area development. BREEAM Communities, LEED for Neighborhood Development and CASBEE for Urban Development all have a special definition of the assessment scale or boundary. So then, based on the study and analysis of Chinese land ownership, urban planning system and residential development process, the assessment scale and platform of the suggested Sustainable Residential District Assessment Tool in China were discussed and found. The residential district together with the regulatory planning was finally selected as the best scale and platform of the assessment tool for three reasons. First of all, in Chinese residential development system residential district is an integrated unit both in the function and structure, which means many important sustainable actions or measures over block scale, such as enhancing accessibility of public transport, creating ecosystem networks or improving sustainable community residents structure can be employed on this level. Second, because the regulatory planning has powerful practicality and operability, an assessment tool based on it can integrate and control the scheme, design, construction and management process efficiently, which can ensure that the sustainable target can be accomplished successfully. Third, such an assessment tool on residential district can provide a better and more equitable condition for green building construction to different building development companies in a same area and reduce the low-price low-performance races between them because of its integrated considerations, assessments and requirements. Forth, the residential district scale and regulatory planning platform can make the assessment tool cooperate better with the assessment tool for building scale—the Evaluation Standard for Green Building without any technical conflicts.

The research then tries to make a better understanding of the concept and technical details of the sustainability assessment tools for big scale residential area development. The three existing assessment tools—BREEAM Communities, LEED for Neighborhood Development and CASBEE for Urban Development—were studied, compared and analyzed to establish the framework of assessment items for the suggested rating system.

After the suggested Sustainable Residential District (SRD) Rating System was established, four residential district developments in China were assessed by the suggested sustainability assessment tool as a test. They all have suitable scales, planned, developed and constructed in recent years, and could represent the general performance level on sustainability issues of residential district developments in China. From the feedback of the test assessment cases, the practicability and usefulness of the suggested tool- the Sustainable Residential District (SRD) Rating System is acceptable for the assessment participants. The involved developers, planners thought the SRD Rating System is a very helpful tool which allows them to make a general and clearer view of the residential district projects on the sustainability issues. It is useful in the practice and the difficulty is acceptable. And the hypothetical scale of the rating levels, which is established on the research of existing assessment system, is also considered to be acceptable in the test. In addition, the comparison and analysis of the results of the four test cases indicated the state of residential district development on the sustainability issues in China to some extent. Generally speaking, their performance on the Land use & site, Water and Ecology is much better than other aspects which need more efforts to improve in the future.

6.3 Research Contribution

The innovation and contribution of present research can be expressed as the following aspects.

First, this research executed a systemic analysis on the Chinese land ownership, urban planning system and residential development process and an integrated study on the state of sustainable development and green building in China. The research identified the necessity of sustainability assessment on the area scale and discussed the possibility and benefits of the residential district as the assessment scale.

Second, this research developed a three-level (Categories, Indicators and Parameters) framework of assessment items for the suggested Sustainable Residential District Rating System. The assessment items referred to the existing assessment tools and relevant national standards of China and were determined through the questionnaire of local experts and laymen. In addition, the research developed a weighting system which can define the importance of each assessment category, indicators and parameters according to the local context of China. The Analytic Hierarchy Process (AHP) methodology and the software of Expert Choice were employed by the research to establish the weighting system.

Third, the outcome of the research was the suggested Sustainable Residential District (SRD) Rating System- computer based program- that suits the Chinese context. The practicability and usefulness of the suggested tool was tested by four cases which are all new town residential district projects in Tianjin. And with the suggested tool, an analysis of these residential districts' performance on the sustainability issues was made to understand the general state of the residential district developments to some extent in China.

Finally, present research provided a research framework and methodology which was referential and helpful for other researchers who concentrate on the assessment of sustainability both for the area and the building scales in China.

6.4 Limitations of the research

Besides the contribution, there are also some limitations in present research. First limitation is the research boundary. Present research attempted to establish a suggested Sustainable Residential District (SRD) Rating System (a computer program) that suits all Chinese cities and regions. While, limited by the research period and data obtainable, present research only focuses on the region of Tianjin, both for the involved respondents and the tested cases. Thus, the conclusion drawn from them could not represent all regions and cities of China. Second, for a mature assessment tool four test cases are not enough especially for the rating scale of score percentage. Now, the hypothetical scale of the rating levels was established on the research of three existing assessment systems for the area scale. Though one of the four cases got a rating of Good, there is still an absence of cases which can get the Excellent rating. It is better to apply this suggested rating system to assess more residential district development projects to verify the feasibility. However, it is impossible to completely carry out so much work within a short period. The third limitation lies in the framework of assessment items. They were established based on the study of the three existing assessment tools and reference relevant national standards of China. Their feasibility should be tested with cases of all regions of China. Through the assessment practice, some inappropriate assessment items should be deleted and new more appropriate items from the real practices should be added into the framework. The idea framework of assessment items, together with the weighting system should be determined by local context and experts and differ among regions of China.

6.5 Future direction of the research

The purpose and objectives of this research have been successfully achieved. However, there are still a number of opportunities for the research to develop in the future. The suggested areas are as follows:

First, more data and projects in different regions of China should be investigated. In future, more experts from various regions and cities of China will be involved in the research to adjust the suggested rating system and develop a special sub-version for their local region or city, which will enhance the practicability in the local context.

Second, more residential district development projects from different regions and cities of China will be assessed with the suggested rating system. The results will be analyzed to test the practicability of the system in different contexts and to find the shortage and problems which should be improved and solved in order to mature the rating system.

Third, the assessment tool for other type settlements such as the industrial area and other type building should be explored in future. Now the BREEAM, LEED and CASBEE have all developed a system which consists of special versions for different types of building, settlement and development. Such a system provided a more comprehensive platform which could guide various developments in the city on the sustainability issues. Now in China, there is still absence of such a system. Hence, paying more effort to establish an assessment system which covers the main types of buildings, settlements and developments is critical to ensure the sustainable development in the Chinese cities in the future.

Bibliography 参考文献

[1] ANDO S, ARIMA T, BOGAKI K, et al. Architecture for a sustainable future[M]. Tokyo: Architectural Institute of Japan, 2005.

[2] COLE R. Shared markets: coexisting building environmental assessment methods[J]. Building research & information, 2006, 34(4): 357-371.

[3] GULYAS A, UNGER J, MATZARAKIS A. Assessment of the microclimatic and human comfort conditions in a complex urban environment: modelling and measurements[J]. Building and environment, 2006, 41(2): 1713-1722.

[4] HAAPIO A, VIITANIRMI P. A critical review of building environmental assessment tools[J]. Environmental impact assessment review, 2008, 28(7): 469-482.

[5] HIKMAT H A, SABA F N. Developing a green building assessment tool for developing countries: case of Jordan[J]. Building and environment, 2009, 44(5): 1053-1064.

[6] HOLDEN M, ROSELAND M, FERGUSON K, et al. Seeking urban sustainability on the world stage[J]. Habitat international, 2008, 32: 305-317.

[7] KAATZ E, ROOT D, BOWEN P. Broadening project participation through a modified building sustainability assessment[J]. Building research & information, 2005, 33(5):441-454.

[8] KIBERT C J. Sustainable construction: green building design and delivery[M]. 1st ed. New York: John Wiley & Sons, 2005.

[9] LARSSON N, COLE R. Green building challenge: the development of an idea[J]. Building research & information, 2001, 29(5): 336-345.

[10] LÜZKENDORF T, LORENZ D. Using an integrated performance approach in building assessment tools[J]. Building research & information, 2006, 34(4): 334-356.

[11] MALIENE V, HOWE J, MALYS N. Sustainable communities: affordable housing and socio-economic relations[J]. Local economy, 2008, 23(4): 267-276.

[12] MALINE V, MALYS N. High-quality housing: a key issue in delivering sustainable communities[J]. Building and environment, 2009, 44(2): 426-430.

[13] NIE M S. Technical assessment handbook for ecological residence of China[M]. Beijing: China Architecture & Building Press, 2002.

[14] REIDSMA P, KÖNIG H, FENG S, et al. Methods and tools for integrated assessment of land use policies on sustainable development in developing countries[J]. Land use policy, 2011, 28(3): 604-617.

[15] ROSELAND M. Sustainable community development: integrating environmental, economic, and social objectives[J]. Progress in planning, 2000, 54(2): 73-132.